Prairie Farmer's Poultry Book

How To Make The Farm Poultry Flock Pay

by William Osburn

with an introduction by Jackson Chambers

Introduction

I am pleased to present yet another title on Poultry.

The work is in the Public Domain and is re-printed here in accordance with Federal Laws.

As with all reprinted books of this age that are intended to perfectly reproduce the original edition, considerable pains and effort had to be undertaken to correct fading and sometimes outright damage to existing proofs of this title. At times, this task is quite monumental, requiring an almost total "rebuilding" of some pages from digital proofs of multiple copies. Despite this, imperfections still sometimes exist in the final proof and may detract from the visual appearance of the text.

I hope you enjoy reading this book as much as I enjoyed making it available to readers again.

Jackson Chambers

You'll get 'em by the pailful if you handle your hens Osburn's way

DEDICATED TO FARMERS' WIVES
AND DAUGHTERS, WHOSE DILIGENCE AND
SKILL IN HANDLING THE MANAGEMENT OF THE FARM
FLOCK ARE RESPONSIBLE IN A LARGE MEASURE
FOR THE MAGNITUDE OF THE
POULTRY INDUSTRY

Mr. Osburn at work on Prairie Farmer Poultry Book

Preface

THIS book is written to help farmers make more money from their poultry flocks.

Success with poultry on the farm depends on intelligent care and feeding. Any farmer or farmer's wife, or farm boy or girl, who will study the methods described in this book, and follow them carefully, can make the farm poultry flock a highly profitable enterprise. We have endeavored not only to point out profitable methods, but also to indicate dangers and pitfalls to be avoided.

Any branch of farming is always more interesting if we know the "why" of things. We have therefore devoted some space to a discussion of the scientific principles that lie back of successful poultry methods, in order that the reader may understand the wonderful process by which nature produces eggs and meat through the medium of the hen.

The farm flock represents fully 80 per cent of the poultry industry. Ninety per cent of all farms produce poultry. The value of all chickens and eggs produced in 1919 was $1,047,989,919. These figures do not represent the full magnitude of the industry, for they do not include fowls other than chickens, such as turkeys, ducks and geese, nor poultry produced on small estates. A safe estimate of the annual production of poultry and eggs in the United States is $1,500,000,000.

For this large asset to the nation the farmer and his family deserve great credit. As the farmer himself is usually occupied in the larger and heavier tasks of the farm, the care of the poultry generally falls to the farmer's wife or upon the daughter or son, or possibly some elderly member of the household. This is possible because the task does not require heavy manual labor. Wherever the task falls it means constant attention to minute details and a high degree of knowledge and skill. To help solve some of the problems which will surely present themselves is our aim.

Table of Contents

List of Illustrations

LIST OF ILLUSTRATIONS

PRAIRIE FARMER'S POULTRY BOOK

Chapter I

The Farm Flock

THERE is little danger of overestimating the importance of poultry culture. Compared with other farm interests it has earned a place of deserved recognition. This is proved by the fact that of the 6,448,336 farms in the United States 5,800,000 are engaged in poultry production. And there are good reasons for this recognition.

The flock furnishes a good percentage of the family dietary. It contributes something to every meal, food of high nutritive value. Probably 35 per cent of all poultry products is consumed on the farm. The remaining 65 per cent goes into the channels of trade to feed the world. The larger proportion is consumed locally, but an increasing annual amount is exported, thus adding to the wealth of the nation. It is evident therefore that the farm flock is of no insignificant value, not only as a source of food for the family but as a source of revenue. Sometimes this income furnishes the allowance for the farmer's wife; sometimes it is sufficient to pay all the table expenses of the household.

The farm flock contributes to other farm interests. Fowls destroy vast numbers of insect pests. It is estimated by the U. S. Department of Agriculture that the annual loss to the farmer by reason of insect pests is $1,555,000,000. The annual saving by birds is $444,000,000. Every agency that reduces this loss aids the farmer in his task of production. If hens are allowed to forage in the orchard they will check the ravages of curculios and moths; chickens and turkeys in the meadow or cornfields destroy many larvæ of harmful insects.

The farm flock utilizes the waste, saving much that would otherwise be lost. Fowls are great foragers and gather much of their subsistence from vegetation, the seeds of noxious weeds, from grain gleaned after harvest, and from waste grain in the feed-lots. These would be a total loss were it not for the farm flock.

There is another consideration outside the commercial importance of the industry. It is the appeal to the æsthetic. What in nature is more beautiful than a bird? Blending colors, symmetry of shape, curved outlines and grace of carriage all appeal to the eye and, through the eye, to the nobler senses, thus contributing something to the joy of living and the development of character.

In order to keep the farm flock at a point of maximum production there should be a program of action. Here is an outline of procedure:

1. **Keep a purebred flock.** It will cost no more to feed a purebred than a mongrel. The mongrel will be neglected, but the purebred will command greater interest and receive better care and prove more productive.

2. **Make conditions favorable** for the flock. This means regular care, a dry, well-ventilated building with an abundance of light, and cleanliness in everything—clean water and food in clean vessels for clean fowls in clean buildings upon clean premises.

3. **Select good foundation stock,** strong in vitality and prepotency, and free from hereditary taint or physical deformity, and follow a system of breeding that will perpetuate the good qualities and eliminate the weaknesses.

4. **Plan for early hatching.** The early hatched pullet makes the early layer in the fall and the matured breeder in the spring.

5. **Eliminate the boarders.** Every poultry breeder should qualify to do his own culling. This is especially important in the spring when the majority of the flock is coming into laying condition and in the fall when the "quitters" are easily detected.

6. **Adopt a system of feeding** that will give results. This means a balanced ration in all seasons. It means food elements which contribute to health, growth and high production. It means planning for green feed in winter by planting cabbage, carrots, mangels, alsike or alfalfa in season.

7. **Make a study of enemies and diseases** and conduct a vigorous campaign against them. Enormous losses are avoided by outwitting the enemies. The growing stock must

be protected. The first approach of disease should be noted by watching for abnormal conditions. If disease appears, sick fowls should be isolated promptly and dead fowls, as well as other dead animals on the premises, should be cremated with dispatch.

8. **Store eggs for fall and winter consumption.** When prices are not remunerative, why should the poultry farmer sacrifice? He can store for better prices as well as the grain farmer. His product will be of better quality than eggs which go through the cold storage plants, because packed when perfectly fresh.

9. **Study the problem of marketing.** Culling for market should begin early when prices are good. A broiler will sometimes command a better price than the same fowl held over till the following spring. To market breeding stock and eggs successfully requires discreet advertising.

10. **Eliminate waste and all unnecessary expenses.** This program may appear visionary because so few attain to its requirements. The ideal is important in any enterprise, for it is only by striving for an ideal that we reach any degree of excellence. To show how in a measure this program may be worked out in practice is our aim in future chapters of this book.

Don't Neglect the Poultry

Opportunity is at the door of every enterprise. Welcomed and seized she leads into fields of promise and reward; neglected her invitations and counsels are in vain. Neglect spells loss. A beautiful damsel was encouraged by a good genius to pass through a cornfield once, and only once. She was urged to pluck an ear, large and beautiful, for according to its size and beauty would be its value to her. She passed many ears large, ripe and beautiful, but neglected to pluck, hoping to find one yet more beautiful. As the day declined she reached a portion of the field where the stalks were thin and barren. And, at last, as the evening closed upon her she found herself at the end of the field without having plucked an ear of any kind. Opportunity had flown and with it the promised reward.

Neglect is the cause of nearly all our disappointments in poultry raising; it is responsible for nearly all the leaks and losses which end in failures. Watch the following common leaks and losses that often keep farm flocks from being profitable:

Loss in purchasing. Neglect to secure good foundation stock is a prime cause of failure. To make a beginning with weaklings where physical vigor has been undermined with disease is equivalent to making no beginning at all. It would be better to pay $100 if necessary for a pen of five birds. physically sound, and that meet standard requirements, and to build upon this foundation, than to spend the same money for a hundred specimens devoid of vigor and representing careless breeding.

The same principle applies to the purchase of equipment and feeds and the construction of buildings. It is true that many poultry appliances can be made at home and will give good service, but the things that must be purchased should have quality and durability. Moldy feeds may be cheap but in the end they will mean the loss of the flock, or, at least, the loss of profits. The poultry building need not be expensive but should be well built, sanitary and convenient.

Loss from improper feeding. Excessive feeding may cause intestinal trouble, liver disease or apoplexy, resulting in the loss of many fowls. The lack of a well-balanced ration means a loss in egg production and retarded growth of the edible carcass, and these are the main sources of income. Wasteful feeding often cuts a hole in the profits.

Loss from careless handling of eggs. Let us follow an egg from producer to consumer. It may be cracked at the outset because the nest is not provided with suitable material. It may be overheated, causing incubation and subsequent decomposition. It may be frozen, cracking the shell and producing a so-called leaker. It may be rough handled, breaking up the contents, producing a watery consistency. The same deterioration may happen in the hands of the country merchant or in transportation to the wholesaler. If it goes to cold storage it may be kept in storage too long and become stale. When it goes to the retailer it still further may be subjected to too much cold or heat or dampness or rough handling resulting in more deterioration.

On the table of the consumer it may have a small percentage of the value it had when fresh laid on the farm. The losses on a 30 dozen case of eggs, due to heat and dampness, run from five cents to $3; due to freezing, from 10 cents to 60 cents per case; due to rough handling, from 5 cents to 25 cents. This was the estimate of the food administration during the world war. From producer to consumer the total annual loss of the egg crop is estimated by the government at $50,000,000.

Leaks due to incorrect incubation. Many thousand of eggs should never go to the sitting hen or incubator, because defective in size, shape and contents. Many contain weak germs because of weak foundation stock or improper feeding. Many are destroyed by neglect of the hen or careless handling. A good average hatch is counted at 50 per cent of the eggs set. Here is a 50 per cent loss that to large extent can be eliminated by scientific breeding and more careful management.

Losses due to enemies. Their name is legion. They populate air and water, they teem in untold millions in the soil, they congregate upon the surface of the earth in great armies of destruction; they stand at the gateways of life and gloat over their prospective victims.

> "A constant watch they keep;
> They never slumber, never sleep,
> Lest they should lose their prey."

The baby chick, emerging from its shell, looks out upon a world of living things in innocence, but is marked for destruction, for there is a constant warfare of life upon life.

It is the aim of every poultry producer to bring to maturity not less than 50 per cent of the chicks hatched. This goal is seldom reached. Here is a 50 per cent loss that may be averted in large measure by eternal vigilance and persistent warfare against the foes of the flock.

Magnitude of Industry

Here are some figures showing the magnitude of the poultry industry. The following table shows the number of fowls of all kinds on farms, January 1, 1920, also an estimate of all fowls not on farms, but on the back lots of towns and cities and on small estates of less than five acres. It also shows the value of these fowls and the value of all fowls and eggs produced in 1919. As the government does not collect statistics of poultry on back lots and small estates, the figures given are merely estimates and are indicated by the figures in black face.

Table No. 1.—Poultry in United States.

Items Jan. 1, 1920	Total number on farms and small estates	Value of poultry on farms and small estates	Value of fowls and eggs produced—1919
Chickens	359,537,385	$349,508,867	$1,047,989,919
Turkeys	3,627,028	12,904,989	38,714,967
Ducks	2,817,624	3,373,966	10,121,898
Geese	2,939,203	5,428,806	16,286,418
Guinea Fowls	2,410,421	1,582,313	4,746,939
Other Fowls	1,493,861	595,116	1,758,348
On small estates	93,206,316	73,970,040	221,910,120
Totals, all fowls	466,031,838	$447,364,097	$1,341,528,609

The number of chickens produced in 1919 was 473,923,935 and the number of eggs produced in that year was 1,656,267,200 dozens. In 1909 the Secretary of Agriculture estimated the annual income from poultry products at $750,000,000, and that was equal to the combined value of all the gold, silver, iron, and coal mined in that year. According to the table given above the total value of all poultry products in 1919 is estimated at $1,340,000,000. This is equal to all the oats, barley, rye, buckwheat, flax, and rice produced in that year. This large income is a great asset to the nation for which the farmer receives due credit and full measure of reward.

The value of the industry, measured in money, is not the highest consideration. The food value to the nation is even more important. Consider the quality of the food product. What is more appetizing and nutritious than the flesh of fowl? And the egg, the universal article of diet, finding its way into every home, is the great tissue builder. It supplies the needs of the brain worker as well as the manual laborer. It furnishes the vitamines so much needed by growing children—the growth principles so essential to health and physical development.

If all the farms should cease in poultry production, or if for any cause all the hens should cease to lay, it would be a national calamity.

Chapter II

The Factory and Workmen

POULTRY production may be likened to the operation of a great manufacturing concern in which there are the factory, the workmen, the raw material and the finished product. The factory is represented by the living bird, the workmen are energized cells and tissues of its organism, the raw material is the food and other materials which are transformed into poultry products, and the finished product is represented in flesh, eggs and feathers.

The Factory

Our first concern is a study of the factory, its systems of machinery and the work to be accomplished by each. The common hen is our illustration for she is queen of the poultry world. Around her revolve the chief interests of economic importance.

EXTERNAL PARTS. The external structures of a fowl are:

1. **Head,** consisting of the beak, comb, face, eyes, ears, ear-lobes and wattles.
2. **Beak,** consisting of upper and lower mandibles.
3. **Nostrils,** located in the upper mandible.
4. **Comb,** which may be single, rose, pea, V-shaped or strawberry.
5. **Face,** naked side of head.
6. **Eyes,** the color being determined by the iris.
7. **Wattles,** pendulous membranes beneath the lower mandible.
8. **Ear,** behind the eye and covered with tuft of feathers.
9. **Neck,** consisting of front and cape. The cape is called the hackle in the male and comprises the feathers of lower neck resting upon the shoulders.
10. **Breast,** feathers covering breast bone.
11. **Shoulder,** feathers covering base of wing.
12. **Wing-bow,** feathers covering side of wing.
13. **Secondaries,** or wing-bay, large wing shafts above the primaries.
14. **Primaries,** lower flight feathers.
15. **Primary coverts,** smaller feathers covering flights.
16. **Secondary coverts,** covering base of secondaries.
17. **Back and sweep,** the latter term applying to feathers over base of tail, called the saddle in the male.
18. **Cushion,** feathers on each side of base of tail.

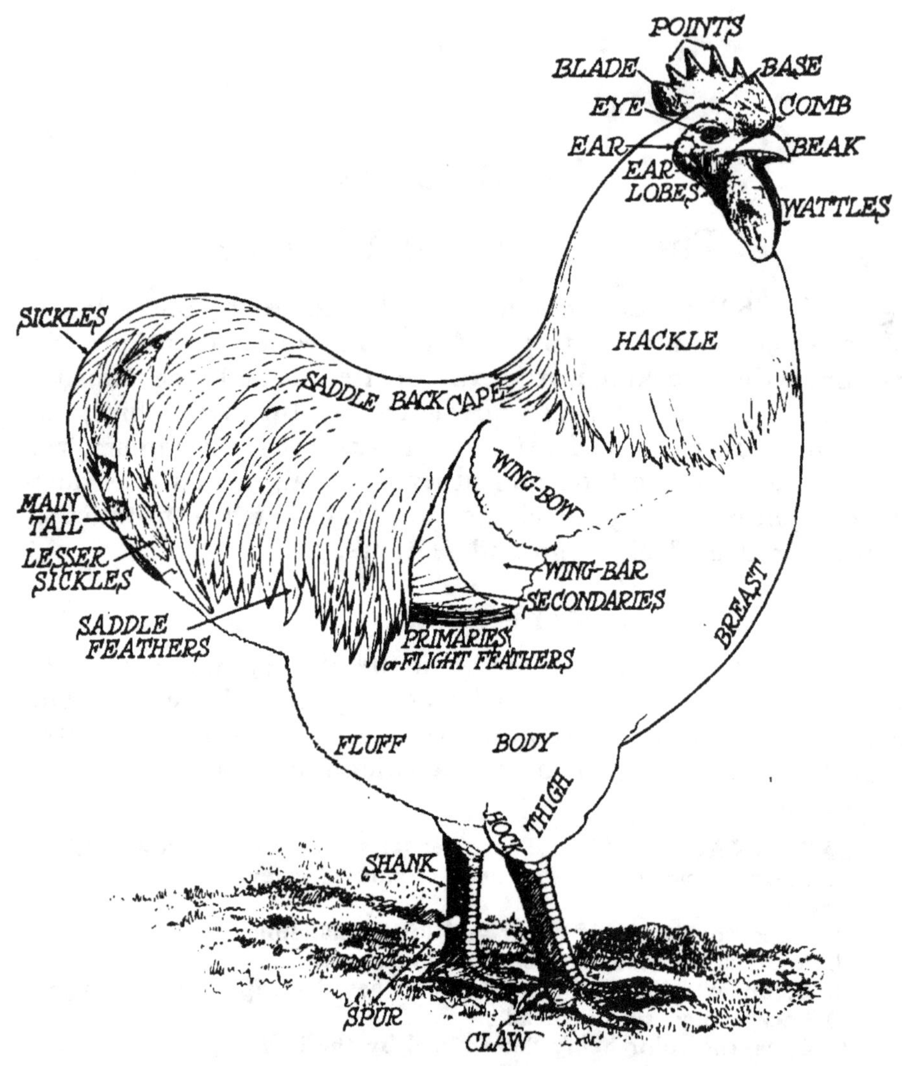

19. **Saddle,** a term applied to flowing feathers over base of tail in male.
20. **Tail,** including the main tail feathers and the sickles, the long curved feathers covering the tail shafts.
21. **Tail coverts.**
22. **The fluff,** region below the vent.
23. **Body,** all between back and sternum.
24. **Thigh,** the leg above the shank.
25. **Hock,** knee joint, or joint between shank and thigh.
26. **Shank,** between hock and toes.
27. **Spur,** a horny growth on side of shank.
28. **Feet,** including ball of foot, toes and nails.
29. **Toes,** usually four, may be five in number, and may be feathered or naked.

A knowledge of all these sections of a fowl is necessary to comprehend the descriptions of purebred specimens. They perform very important functions and are designed for protection, locomotion, obtaining food, etc.

Internal Systems and Structures

In a study of the internal structures of a hen we find nine important systems of organs, each having important work to accomplish. They are: Skeleton, respiratory, circulatory, digestive, excretory, reproductive, muscular, nervous, and tegumentary.

The Skeleton

The skeleton is the bony framework of the body. A bone consists of the periosteum, the white fibrous covering that supplies nourishment; bone-proper, consisting of the carbonate of lime and phosphate of lime; and the marrow, or central soft tissue, whose chief work is to help in making red blood cells.

The parts of the skeleton are:

SKULL:

Mandibles, upper and lower jaws.
Cranium, bony box containing the brain.

NECK:

13 vertebrae, the one next the cranium being called the atlas.

TRUNK:

Dorsum, or back, comprising the seven **thoracic vertebrae** to which the seven pairs of ribs are attached.

Sacrum, a term used to comprise all the vertebrae between the thoracic and the caudal vertebrae. They are fused together so as to form one solid structure.

Caudal Vertebrae, six in number, which support the structures of the tail.

Pelvis, comprising three distinct bones on each side, known as the ilium, ischium and pubis. These form a protection for the kidneys and other viscera. They unite to form a socket for the femur, and they are fused to the sacrum so as to form a continuous arch. The points of the pubis, which can be located just below and on each side of the vent are known as the pubic bones, or so-called "lay-bones."

Shoulder circle, comprising:

Scapula, or shoulder blade.

Caracoids, strong bones extending from sternum to shoulder. They form a point of attachment for the humerus and the hold the sternum and shoulders apart.

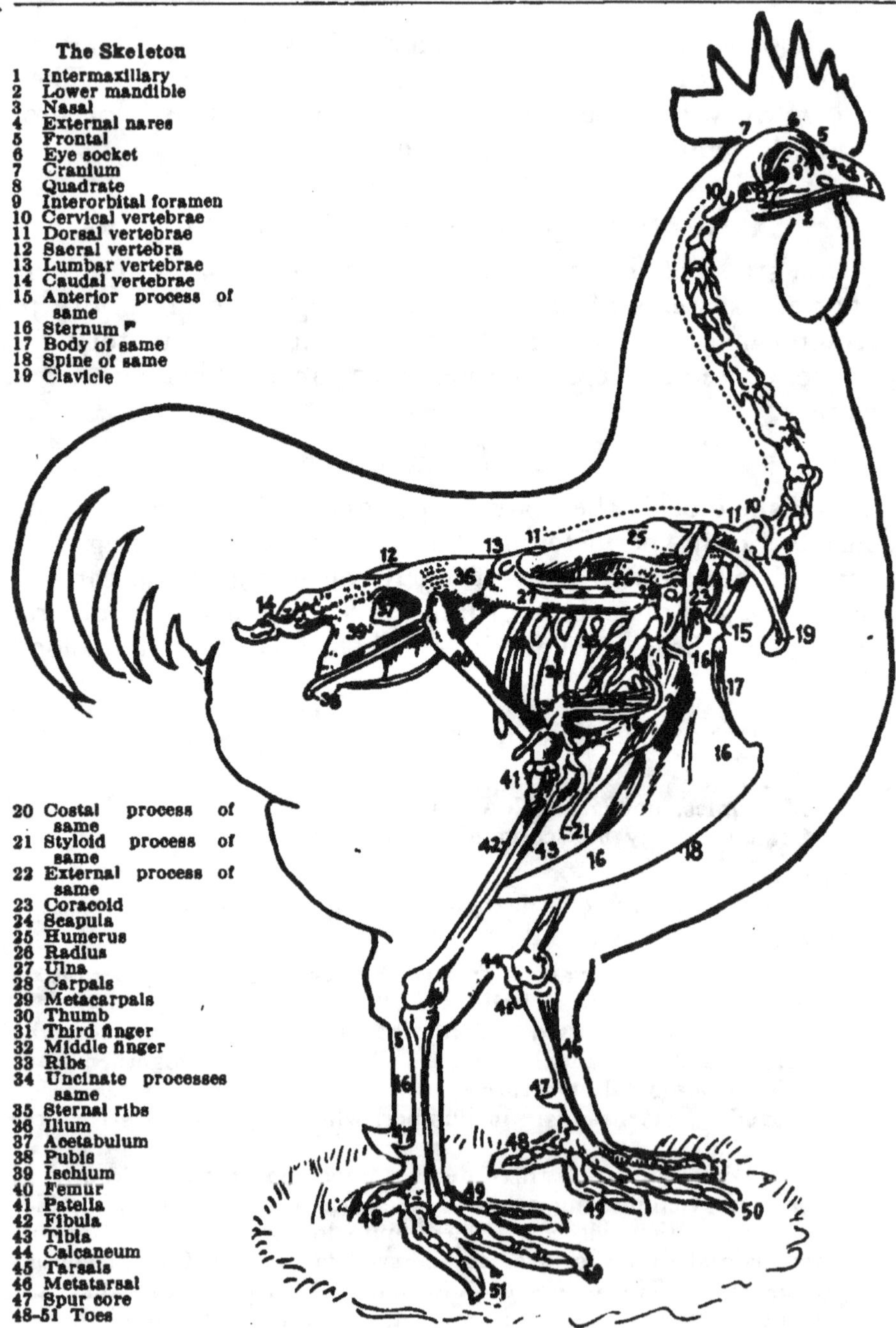

Clavicles, one on each side, which unite to form the so-called "wish-bone," or "merry-thought."

Sternum, or breast bone, a boat-shaped structure, forming the floor of the vital organs. The sharp, bony edge of the sternum on the lower side is known as the keel, and to this the muscles of the breast are attached.

LIMBS:

Wings:

Humerus, or upper arm, attached to the shoulder girdle.

Fore-arm, comprising ulna and radius.

Hand, comprising two small bones united at their ends.

Fingers, the thumb being attached to the upper end of arm and the two remaining fingers to the lower end.

Legs:

Femur, or thigh-bone, attached to the pelvis.

Lower leg, consisting of the tibia and fibula, the former, large, and the latter small and splint like.

Shank, or tarso-metarsus, attached to the lower leg at the knee joint, or hock, and bearing the **spur.**

Toes:

Inner toe, or hind toe, which has two joints.

Second toe, or inner, front toe, having three joints.

Middle front toe, having four joints.

Outer front toe, which has five joints.

A knowledge of the structures of the skeleton will be of value in mating and culling and in preparing the carcass for market as well as dissecting for table use.

The Respiratory System

Respiration in fowls includes the acts of receiving air into the lungs (inspiration) and expelling air from the same (expiration). These processes are accomplished by muscular action, raising and lowering the sternum. By the respiratory system oxygen is conveyed to the blood and vapor of water and waste matter thrown off from the body. As the fowl has no sweat glands to eliminate vapor of water and certain waste matter which accumulates in the blood by the process of oxidation, the respiratory system accomplishes this work to large extent. Respiration in man is 18 times per minute, but in the fowl it is more rapid, or 33 times per minute.

The organs and functions of the respiratory system are:

1. **Nostrils,** through which the air is conveyed to the pharynx.
2. **Pharynx,** or throat.
3. **Larynx,** the enlarged cartilaginous, or gristly, opening into the windpipe, or trachea. It modifies the voice, which is produced at the lower end of the trachea.
4. **Trachea,** or windpipe, the cartilaginous tube formed by rings of gristle which conveys the air to the lungs.
5. **The bronchi,** or bronchial tubes, branch from the trachea and enter the lungs. Some pass through the lungs into the large air sacs in the body. Some have blind endings. The small tubes which branch from the bronchi are lined with blood vessels which absorb oxygen.
6. **Air sacs,** spaces outside of the lungs which receive the air from the bronchial tubes. They are large and lined with thin membranes

which contain blood vessels through the walls of which oxygen is taken up. Air spaces also occur in the feathers, bones, and elsewhere, making the bird's body light and buoyant.

A knowledge of the respiratory system is important on account of its bearing upon the hygienic care of the flock and a comprehension of the diseases which attack its organs. The more common diseases of the respiratory system are: Catarrh, roup, diptheria, gapes, pharyngitis, bronchitis, aspergillosis, inflammation of lungs and, occasionally, tuberculosis.

Circulatory System

There are two systems of circulation, the blood vessels and the lymphatics.

Blood circulation. The blood is the red fluid which is the carrier of the food elements to the cells and tissues of the body for their repair and upbuilding and of the waste matter to the lungs and kidneys to be eliminated. About 90 per cent of the blood is water; the remaining portion comprises the corpuscles and the plasma.

The corpuscles are red and white. The red corpuscles in a fowl are nucleated, elliptical discs; in human blood they are circular discs, non-nucleated, and smaller. The color of the red corpuscle is due to the hæmoglobin, containing iron and manganese. The affinity of the hæmoglobin for oxygen results in oxyhæmoglobin, and this becomes the carrier of oxygen from the lungs to the cells of the body. The white corpuscles are nucleated and consist of living protoplasm, exhibiting amœboid movements. They perform important functions, destroying disease germs, healing wounds and building up tissues.

The organs of circulation are the heart, arteries, capillaries and veins. The contraction of the heart forces the arterial, bright red blood coming from the lungs through the arteries to the capillaries, and the veins collect the venous, dark red blood from the capillaries and convey it back to the heart, whence it is forced to the lungs to be again charged with oxygen and returned to the heart for another journey. The heart is a muscular organ with four cavities—two auricles and two ventricles. The pulsation of the heart is more rapid than in other animals, hence oxidation in the lungs and capillaries is undoubtedly more rapid and the blood is, therefore, hotter.

A temperature test of a number of chickens showed a temperature of 101°. The average for man is 98.6°. In a chicken the heart beat averages 150 per minute; in man it is 72 per minute. This explains why respiration is more rapid in fowls than in many other animals and why they soon get out of condition if kept in stuffy, ill-ventilated quarters. The average respiration per minute of a number of chickens was found to be 33. Human respiration averages about 16 times per minute. Several diseases attack the circulatory system of fowls and the blood, such as dropsy, inflammation and enlargement of the heart; thrombosis; cholera; anæmia, infectious leukæmia; and sleeping disease.

The **lymph** is a colorless fluid of value to the blood and originates in the region of the capillaries, being an exudate of serum from the blood into the intercellular spaces. It is collected in very small tubes (lymph capillaries) which convey it to two main vessels, one on each side of the spine, thence upward to the base of the neck, where it is emptied into the general circulation. The lymph vessels in the intestines are called lacteals on account of the whitish color of the lymph fluid, known as chyle.

The Digestive System

The digestive system, as its name suggests, receives the crude food, grinds it, and prepares it for absorption into the blood and for assimilation. It is the great workshop of the factory, working over the raw material for replenishing old cells and for the manufacture of new.

The organs of the digestive system are:

1. **Pharynx,** or throat, which receives the food from the beak and mouth and forces it into the œsophagus.

2. **Œsophagus,** or gullet, an elongate tube capable of vermicular muscular action by which the food is forced downward into the crop.

3. **Crop,** a dilation of the œsophagus, where the food is softened and held in reserve for the further processes of digestion.

4. **Proventriculus,** or stomach. This is the enlarged pouch to which the food passes from the crop. It is provided with glands which secrete the gastric juice, a digestive fluid whose action is to change the food into a condition known as chyme.

5. **Gizzard,** a receptacle of the food as it passes from the proventriculus. The gizzard is provided with a tough inner membrane and powerful muscular walls which assist in grinding the food and mixing it with the digestive fluid, so that before leaving it is reduced to a

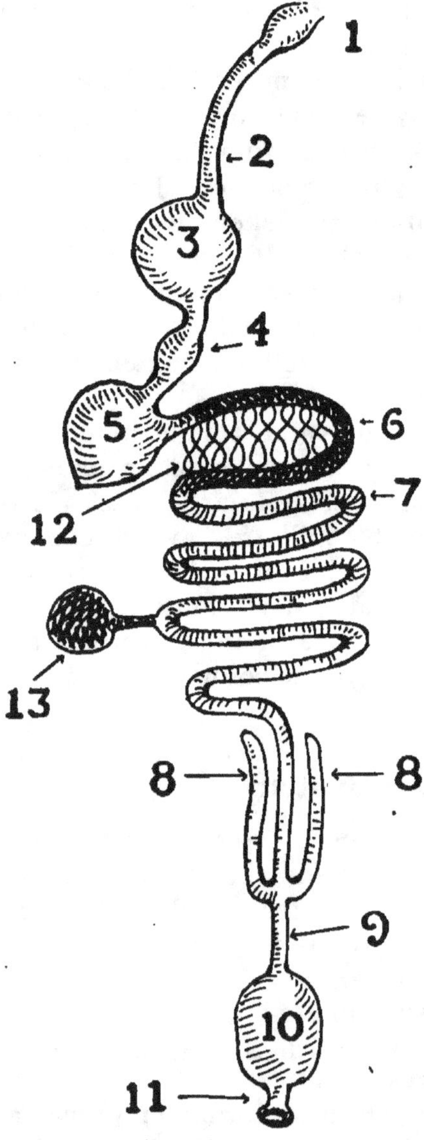

pasty mass. The **gizzard** therefore, is the organ of **mastication**, and the grit which it contains serves as teeth.

6. **Duodenum,** the upper intestine. It forms a curve in the shape of the letter U, between the arms of which lies the pancreas. The food passes from the gizzard to the duodenum where it mixes with the fluid from the sweetbread and the bile from the liver. These fluids, acting with the intestinal juices, change the fatty foods into a condition known as chyle. The chyle is absorbed by the lacteals, carried by the lymphatic system to the base of the neck and emptied into the general circulation. Other portions of the food are absorbed by the capillaries and carried by the portal circulation to the liver, there to undergo certain changes preparatory to assimilation. Protein is changed to peptone by the pepsin of the gastric juice, the peptone being soluble and capable of absorption through the intestinal walls into the circulation; carbohydrates, consisting of sugar and starches, are changed into glucose, which is also soluble and capable of absorption.

7. **Small intestine,** or ileum, which receives the food from the duodenum. It lies in folds and convolutions which are connected and held together by a thin membrane known as the **mesentery.** At the lower end of the small intestine two branches are thrown off, known as **ceca.** These extend forward, parallel with each other, and are closed at their upper extremities, hence are called blind pouches. The functions of the small intestine are to carry on the work of digestion and complete the work of absorption. To this end it secretes a digestive fluid, known as the intestinal juice, and is covered with elevations known as villi to increase the absorbing surface.

8. **Ceca.**

9. **Large intestine,** or colon, which begins where the ceca branch off and is a straight tube to the cloaca. Its function is to convey the undigested portions of the food, or fecal matter, to the cloaca.

10. **Cloaca,** an enlarged pouch at the end of the large intestine. This receives the waste matter from the raw material and urates from the kidneys and discharges them through the vent.

11. **Vent.**

12. **Pancreas,** as stated, lies in the fold of the duodenum. It secretes the pancreatic juice, a digestive fluid which has three ferments, each of which performs an important office in digestion. The trypsin changes the albumin to peptone, the amylopsin changes starch to glucose, and the steapsin acts upon the fat to emulsify it.

13. **Unabsorbed yolk.**

Other digestive organs are:

The liver, a large, soft, glandular organ lying between the heart and gizzard. It is a very important organ of digestion. It acts upon the peptone which comes to it from the intestines through the portal circulation. The peptone is converted back to albumin and thus prepared to become a constituent part of the blood and ready for assimilation by the cells throughout the body. The liver also changes the glucose into glycogen, which is taken up by the blood and is oxidized as it is carried onward in the circulation, thus giving heat and energy to the body. The liver also performs an important function in destroying disease germs and eliminating poisons which may come to it through the portal circulation. The liver secretes the bile, which is collected in a sac known as the gall sac, whence it is conveyed by the bile duct to the duodenum. The bile plays an important role in digestion, for it not only lubricates the walls of the intestine but aids the other digestive fluids in performing their work.

Thus we see that the digestive system is a marvelous piece of machinery. It is concerned with softening, grinding, dissolving, and digesting by certain chemical changes, absorbing and assimilating the raw material. A knowledge of its organs, their mode of work, and their functions, is very vital to the poultry keeper. If these organs function properly, are provided with the elements needed for the up-building of the body, as well as their own recuperation, all goes well—the fowl is healthy and production continues. Otherwise there are disturbances throughout the factory, and fatal diseases may follow. The whole problem of feeding and maintaining health in the flock hangs very largely upon this knowledge.

The Reproductive System

This system is concerned in the perpetuation of the species. It produces the egg which contains the living primordial cell from which the new individual is to spring.

The organs of the reproductive system are the ovary, oviduct and the cloaca.

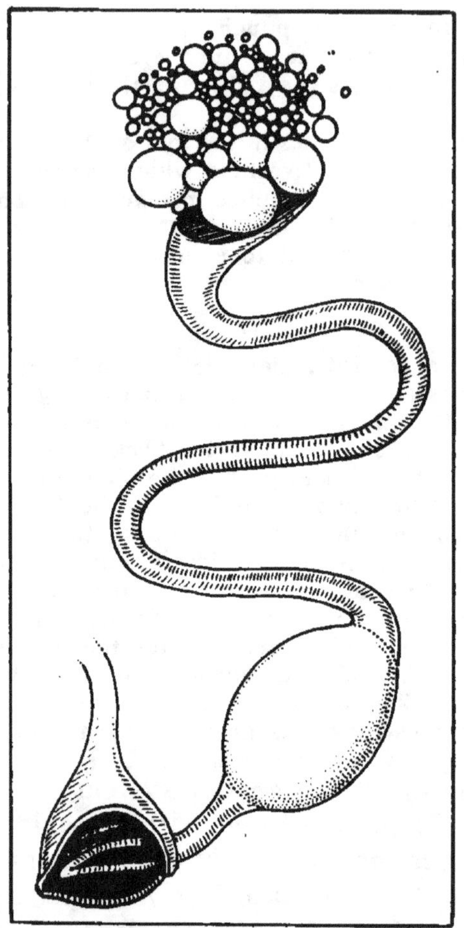

Egg organs showing ovary, oviduct and cloaca. **The expanded opening into the oviduct is the infundibulum.**

The ovary comprises a cluster of spherical bodies, oöcytes, which lie to the left side of the back, just beneath the spine. These bodies are at first but microscopic points, but they are living, protoplasmic cells. They develop into the yolks and are surrounded by a delicate membrane, known as the **ovisac,** or follicle. When the yolk becomes ripe, the follicle cleaves and allows the yolk to escape into the oviduct.

The **oviduct** is a convoluted tube, about 18 inches in length. The funnel-like mouth of the oviduct is called the **infundibulum.** As the yolk is conveyed downward through the oviduct it receives, in the upper portion, the albumen; in the central portion the membranes surrounding the egg are formed; and, in the lower end, lime is secreted to form the shell. Soon after the completed egg passes into the **cloaca** it is covered with a mucous deposit, or film, which serves to prevent the evaporation of the egg contents and also, in some degree, prevents the access of germs and harmful substances from without. The muscular walls of the cloaca are used to expel the egg.

Male Reproductive Organs

The principal male organs concerned in reproduction are the testes and the seminal vesicles.

The testes lie in the same relative position as the ovary in the female and are opposite the last two ribs on each side. They produce the semen, which consists of the **seminal fluid** and the **spermatazoa,** or sperm cells. **The seminal vesicles** are the tubes that convey the semen to the cloaca. The sperm cells are living cells, oblong in shape, and are provided with whip-like flagellæ, or lashes, by which they are able to swim from place to place.

The reproductive organs, especially those of the female, are subject to a number of abnormalities and diseases, which will be considered in the treatment of diseases. A knowledge of these organs and their workings has a very practical bearing on the management of the flock.

The Nervous System

There are two systems of nerves in a fowl, the **cerebro-spinal** and the **sympathetic.**

The cerebro-spinal system comprises the brain, spinal cord and motory and sensory nerves. This system is the medium of such mental operations as the bird possesses and presides over the senses. The senses of sight and hearing are very acute with fowls, much more so than with human beings. The senses of taste, smell and touch are much more limited.

The sympathetic system consists of a series of ganglia with radiating nerves and it is connected up very intimately with the cerebro-spinal system, and reacts on the digestive, respiratory, circulatory, and reproductive systems.

The operation of every piece of machinery in the factory depends upon the stimulation of the nerves. They are not often affected, but such diseases as apoplexy and epilepsy occur occasionally. Sometimes the nervous system is attacked by worms and other parasites which produce peculiar disorders, probably due to malnutrition caused by the parasites, or to toxic poisoning.

Other Organs and Tissues

The spleen. This is a small, dark red organ which lies above the liver and is attached to the proventriculus. It is not provided with a system of blood vessels and capillaries as in other organs. It is not a digestive organ but is considered the factory of the red blood corpuscles, for the blood, after passing through it, is richer in these elements than when it entered. Red corpuscles are also manufactured in the red marrow of the bones.

The muscular system. Muscles by their contraction and relaxation control the motions of the body. They are composed of fibers, which are striated in the case of voluntary muscles and non-striated, or smooth, in all involuntary muscles except the œsophagus and heart. The peristaltic movements of the alimentary tract, the pulsation of the heart, and the motions of swimming, flying and walking all depend upon this wonderful system of muscles. It is estimated that there are 12,000 muscles in a goose, including muscles that control the movements of the feathers.

The excretory system. Excretion is accomplished by the lungs, intestines and kidneys. The uric acid is eliminated from the blood by the kidneys in the form of urates. These are semi-solid and are conveyed to the cloaca by the urinary tubules. The urates are excreted with the droppings, and can be observed as a white layer constituting about one-third of the excreta. The kidneys are located in the back in the cavities of the sacral region and can be recognized as reddish, granular masses. They often become diseased, and their failure to function results in such serious disorders as gout, rheumatism and uræmic poisoning.

Tegumentary System, or Skin. The skin is the outer covering of the body consisting of several layers of cells which serve as a protection to the bird. The appendages of the skin, such as feathers, spurs and nails are modified scales. The continuation of the skin within the body is known as the mucous membrane.

A Wonderful Machine

We have considered the wonderful machinery of the hen's organism. Every system of its machinery has its special organs and every organ has its special work to perform. It is only when all these parts work in harmony that there can be a healthy and productive fowl.

If by any accident or disease any organ is impaired or put out of action it may seriously affect the working of the whole factory and limit, if not totally curtail, the output of finished products. The importance of understanding these structures is very evident.

A Practical Lesson

To show how this knowledge is of practical application in the care of poultry, one illustration is given. Protein is recognized as a very necessary constituent of the food. This nitrogenous food is acted upon by the digestive fluids and changed into peptone. Why? Because peptone is soluble and readily transfuses through the walls of the intestines. But peptone is a poison to the system, therefore it is carried by the portal circulation to the liver, which has the power to change it to albumin, which is at once assimilated by the blood, and this carries it to the cells to be used by them in building tissues.

If there is an excess of protein taken in the food and an excess of peptone formed, then the liver is overworked. Some of the peptone may go into the circulation to poison the system. The liver becomes congested and there may follow a train of liver diseases and gout or rheumatism, causing the loss of the fowl. Further, uric acid is a nitrogenous compound and is produced in excessive quantity when there is an excess of protein in the food, as often happens in the liberal use of tankage.

As the uric acid is eliminated by the kidneys, they are overworked with resulting congestion and disease. Thus we learn the importance of a balanced ration in which there is a due proportion of protein and carbohydrate. Many illustrations of this nature may be drawn from the digestive system alone.

The Workmen

What are the workmen in the great factory? It is true that an organ of the animal body may function as a unit to perform a definite work, as, for example, the eye performs the function of sight; but the eye is made up of tissues, such as epithelial, muscular, connective and nervous, and each of these has a definite work to do. The tissues also are made up of cells and each cell plays a part in accomplishing the desired end. If the cells should break down there would be failure all along the line. The cells, therefore, are the real workers, and their functioning makes possible the organism, for cells make tissues, tissues make organs, and organs make up the organism, or living bird. The cell is the working unit, just as is the individual in the industrial concern.

Chapter III

The Raw Material

FOOD is any substance which taken into the animal organism contributes to its growth and maintenance. Food is fuel for the engine and raw material for the factory. The term "feed" has a similar scope but is applied more particularly to animals, not to man. The term feed may also apply to a definite portion of food, as a feed of corn or oats.

A ration is a definite combination of foods or feeds.

A balanced ration is one prepared to meet the full needs of the animal organism. The ration to be used depends upon the end desired and the environment of the flock. The ration in summer with free range should be quite different from that of winter in confinement. The ration to force egg production would differ somewhat from a fattening ration. The ration during the moulting season should be adapted to the needs of the fowl in growing new feathers. The baby chick ration should meet the demands of the growing body.

Available Poultry Feeds

Wheat is the most desirable of all grains for poultry. It is more nearly a balanced ration than any other grain. It is free from an indigestible hull. As a feed for growing chicks and laying hens it is of great value. Even though at times expensive, the results obtained warrant its use.

Corn is rich in carbohydrate and, therefore, is an excellent winter feed. Chickens will eat it in preference to any other grain. It is not a balanced ration and, if fed alone, will prove disappointing unless the fowls have free range and have opportunity to secure green feed and insects. It is valuable as a fattening ration and as a part of developing and laying rations.

Oats make a very valuable feed for poultry, but on account of the thick hull with its indigestible crude fiber it is not relished like corn and wheat. It is equal to wheat in protein content and about equal to corn in fat. Its nitrogen-free extract is somewhat less than either corn or wheat. Its ash content surpasses both of these grains. The hull is often provided with sharp beards which irritate the digestive tract and sometimes cause serious trouble.

Clipped oats are free from these objections and, where practical, should be used in the grain ration. Hulled oats would be even better. They can be obtained at the mills in 500 pound lots at reasonable prices. Whole oats should constitute about one-fifth of the grain ration. If fed in the same amount as wheat or corn much of it will be left in the litter to become contaminated and moldy and thus the cause of disease in the flock. Ground oats make a valuable addition to the dry mash. The hulls are objectionable, but the mash is generally picked over and the hulls discarded by the fowls, so that this objection is not so serious. Only heavy oats with thin hulls should be used. Light oats with thick hulls will be rejected by the fowls unless they are on a starvation diet. Scalded and sprouted oats are desirable forms in which to feed this grain.

Rye is not relished by poultry nor do they seem to thrive upon it. The ergot of rye is a dangerous poison.

Barley is a good substitute for corn, having about the same composition. It may be fed with corn to give variety. The hull is objectionable.

Buckwheat has a thick, indigestible hull and otherwise is objectionable. A small quantity in the grain ration will add variety, but otherwise is of little value.

Milo maize, kaffir corn, sorghum seed, etc., are good substitutes for indian corn and are relished by the flock.

Millet seed adds variety when thrown in the litter with other grains and stimulates exercise, but it is indigestible and even injurious to growing chicks.

Ingredients for the Dry Mash

Bran is the outer layer of the wheat kernel and is rich in ash, protein, and fat. It furnishes a narrow ration on account of its comparatively small amount of carbohydrate. The main objection to bran is its crude fiber which is 9 per cent of its composition. This objection disappears when it is combined with meals having a limited amount of crude fiber. On account of its bulky nature it is valuable to mix with the denser meals, and on account of its coarse nature it has a stimulating effect on the mucous membrane and is counted slightly laxative.

Wheat middlings are produced from the layer of the wheat kernel next to the bran. They are richer in carbohydrates than bran and contain less ash and crude fiber.

Corn meal is ground corn and is of the same composition as whole corn. Gluten meal is the portion of the kernel just under the hull. Gluten feed consists of the corn bran and gluten meal mixed. Corn bran is the ground hulls.

Ground oats are whole oats ground fine. They contain the hulls and should therefore be made of heavy oats with thin hulls. Oat middlings, or oat flour, comprise the ground kernels after the hulls are removed. Oat meal, or rolled oats, is the hulled oats rolled out and prepared for human consumption.

Linseed meal is made from flaxseed. It has high percentages of protein and fat and is a valuable feed for poultry. On account of its very laxative nature it should be fed in limited quantities. About 5 per cent in the mash ration will be found advantageous. It is especially valuable in stimulating the growth of feathers.

Animal Feeds for the Mash

Meals of animal origin are of great value in stimulating egg-production. They promote growth in young stock and help adult fowls to lay on flesh. Among those more commonly used are the following:

Meat scrap contains about 50 per cent protein. It consists of meat trimmings, steam-cooked and pressed to sterilize and remove fat, and then ground. **Meat crisps** are made from lean meat and contain 75 cent protein. **Meat meal** is the same as meat scrap only ground finer. Meat crisps, ground fine, are very valuable in the mash for growing chicks. Poor grades of meat scrap are unsafe as food for poultry. Purchase your supply from a reliable house and always test before using. Open the larger pieces and note whether mold is present. Pour boiling water upon a small quanity to discover whether the odor remains sweet. Also examine for hair and pieces of horn, which should not be present.

Tankage is made of poorer grades of material and unless prepared by a reliable house with guaranteed analysis should not be used. High grade tankage is used very extensively.

but on account of its high protein content should be used in limited quantity. About 10 per cent of the mash mixture is advised. Used in excess it is liable to produce gout, rheumatism and other affections.

Fish scrap, prepared from dried fish, is valuable as a proteid food and is used extensively in commercially prepared feeds. It frequently imparts a fishy flavor to the flesh and eggs.

Milk in some form is very valuable for poultry. It may be fed sour or sweet. Granulated milk and dried buttermilk are convenient forms to use when skim milk is not accessible.

Other Constituents of the Mash

Charcoal is a corrective and aids digestion. It should be used in every dry mash to keep it sweet and dry, and should be kept before the fowls constantly in hoppers.

Salt stimulates the secretions and aids digestion. Used in excess, it brings on bowel trouble and often acts as a poison, producing death. Used sparingly, it is of great value in a dry mash. About one-half pound in one hundred pounds of mash is the correct quantity. Dry, finely granulated table salt, free from lumps, is the kind to use.

Ash comprises the mineral salts such as soda, lime, salt, magnesia. The combinations are chiefly chlorides, carbonates, phosphates and sulphates. These substances are usually supplied in foods of vegetable and animal origin.

Fowls also obtain some of the mineral salts from the soil and the water they drink. A fowl given nothing but distilled water and foods containing only pure protein and carbohydrate would soon perish. Phosphate and carbonate of lime are needed to build bone and the shell of the egg. Ash enters into the structure of the feathers and is more or less needed in all the tissues of the body. If the supply in food and water is not sufficient, and this certainly occurs when fowls are kept in confinement, it should be supplied. This is done by feeding oyster shell, soil that is not infested, rock phosphate, ashes, crushed limestone and granulated bone. An adequate supply of mineral salts will prevent fowls from eating their droppings and increase the health of the flock. **Oyster shell** contains carbonate of lime, phosphate of lime and some organic matter. It should be kept before the fowls at all times. Its importance in egg production is shown by the sudden decrease in eggs when the supply runs out.

Granulated bone is valuable not only for the mineral matter it contains but for its protein. If supplied in a separate compartment of the hopper, the poultry keeper will realize its value by the large amount consumed. It is especially valuable as a part of the egg-producing ration.

Water is a mineral. It is a food because it enters into all the tissues and structures of the body. It must be provided or the fowl sickens and dies. It should be supplied fresh daily in clean vessels and is especially needed to soften the food after a full crop is obtained of dry feed.

Green Feeds

Green feeds are greatly relished by poultry. The goose subsists almost entirely upon weeds and grasses during the growing season. The same is true to a large extent with ducks and turkeys. The value of green food in promoting health and increasing egg-production is generally underestimated, and too little effort made to provide it. It contains an abundant supply of ash, and its proteid and carbohydrate nutrients are in easily digested form. When it cannot be supplied in succulent form it should be furnished in dry form. Some of the dry forms are clover meal, alfalfa meal and dried beet pulp.

Alfalfa meal can be purchased at supply houses and many feed mills. If prepared from fresh green hay it makes a valuable substitute for green feed. Its abundant supply of ash and other food elements makes it a desirable meal for the dry mash. As a rule alfalfa is not relished by poultry when fed alone, but as a part of the mash it is in favor, if used in limited quantity, about 10 per cent of the total weight.

Clover meal has about the same value as alfalfa. Alsike clover makes a fine litter for laying hens. The leaves are eaten greedily.

Beets or mangel wurzels make a very desirable succulent green food. They are greatly appreciated by the hens and are eagerly devoured. They may be chopped fine and fed in vessels or cut in slices and nailed to the wall. Beets contain water-soluble C vitamine, but are especially valued for their ash content.

Other succulent feeds are cabbages, carrots, potatoes, turnips, pumpkins, sprouted oats, sprouted rye, swiss chard, lettuce, dandelion leaves, etc. The poultry keeper should plan in spring for his winter supply of succulent feed. A small space devoted to mangels will give a surprising return. Carrots do not freeze readily and make a good feed that can be grown at small cost.

Condiments

We do not advise the use of stimulants and tonics as a practice. When the hens are healthy and happy and are doing full service in filling the egg basket why change a system of care and feeding that has been tried and found successful?

There are times, however, when a tonic will help the fowl to tide over and save it from disease. Frequently the egg organs are dormant and only need a tonic to stimulate them to action in order to bring them into laying condition. Fowls are like human beings, they have their ills and humors and often need a corrective or a tonic, to which they readily respond, though it would be folly to depend upon these alone.

Unless an egg tonic is used in connection with a balanced ration more harm will result than good. If the raw material is not present how can the egg be manufactured? If the nervous system needs a little stimulus to action, or the digestive system is sluggish and needs a tonic to correct abnormal conditions, or the reproductive organs are inactive and need a stimulant to incite the process of egg forming, a tonic may serve a good purpose.

There are a large number of advertised tonics. Some of these are of real value, some are measured by an interrogation point. There are also private tonics which are offered as great secrets and sold for a price. Some of the substances frequently used are cayenne pepper, venetian red, quinine, strychnine, sulphuric acid, ginger, onions, etc. Of these the following deserve mention:

Cayenne pepper is a stimulant to the liver and other digestive organs. It is used to relieve colds, in which case it is given in gelatin capsules. This powerful stimulant should be used in limited quantity, if at all.

Venetian red contains oxide of iron and calcium sulphate. It serves as a tonic to the digestive system and is beneficial to the blood.

Iron, quinine and strychnine are valuable tonics and combined in tablets or capsules will be found a splendid help in restoring to vigor fowls that are off their feed, anæmic or

Of course they'll lay if they have a comfortable, well-ventilated house like this, and are fed the way Mr. Osburn advises

emaciated. Formula No. 1 suggests how the preparation should be made and administered.

Ginger is a tonic very beneficial to all the organs. Combined with other remedies, as suggested in formula No. 2, it provides a tonic of real value.

Mustard is a strong stimulant. Its remedial and tonic character is not fully appreciated.

Onions are valuable as a food and serve as a tonic. They can be fed to growing stock over four weeks old and to adult fowls.

The following tonics and stimulants are recommended:

Formula No. I—Health Tonic

Tonic for colds, asthenia, digestive disorders, anæmia, and general debility due to long continued laying, sitting, or other cause.

Sulphate of quinine	1 grain
Strychnine	1/30 grain
Iron Sulphate	1 grain

Given daily in two-grain capsule until recovery. Mix equal parts of quinine and iron, then fill the capsule with the mixture after first putting in a 1/30th grain tablet of the strychnine.

Formula No. II—Health Tonic

Tonic for indigestion, torpid liver, constipation, diarrhæa, blood affections, and general debility.

Gentian	8 ounces
Mustard	8 ounces
Ginger	4 ounces
Epsom salts	8 ounces
Saltpeter	4 ounces
Iron Sulphate	1 grain

Mix thoroughly and give two tablespoonfuls of the mixture in ten quarts of dry or moist mash, daily until recovery.

Formula No. III—Egg Tonic

Correct feeding and exercise are the best stimulants for the egg organs. Eggs cannot be produced without the raw material and often hens will not lay, though correctly fed, because they do not have sufficient exercise. Sometimes, however, when conditions are apparently just right the flock is sluggish and the hens refuse to lay. Under such conditions an egg tonic may have value. The tonic recommended below has been used to some extent, and those who have tried it have not been disappointed. It is claimed for it that it does not weaken the vitality, if directions are followed, and actually increases the fertility of the eggs. We advise its use simply as a means of stimulating the egg organs to action and to get the hens into the laying mood. It should always be accompanied with a balanced ration and an abundant supply of feed. Directions follow.

To the water or milk for the daily mash add tincture of cantharides, allowing one-fifth of a drop to each hen. A hundred hens would require only 20 drops. A teaspoon contains 60 drops, which would be sufficient for 300 hens. Feed in the moist mash daily for one week. During the second week alternate the tincture of cantharides with black gunpowder, i.e., give the tincture one day and the gunpowder the following day. Add one tablespoonful of the powder to the water or milk for the daily mash. This will be sufficient for one hundred hens.

After the second week feed the tincture of cantharides only once a week, but always follow it on the succeeding day with the black gunpowder. Gunpowder is composed of 70 per cent of niter, 15 per cent of sulphur, and 15 per cent of charcoal. It is sometimes difficult to purchase. Any druggist can prepare it according to the formula given above. Continue this feeding for four weeks and then discontinue.

Caution: Tincture of cantharides is a powerful poison and should be kept out of the reach of children.

Formula No. IV—Egg Tonic

To the mash for 50 hens add 8 teaspoonfuls of mustard. The mixture may be fed dry or moist, the moist mash being recommended. Mustard is a fine tonic and corrective, a splendid stimulant to the egg organs, and promotes the health of the flock. It is claimed that no harmful effects follow its use.

A Word of Caution

In the foregoing discussion we have pointed out the nature, composition and value of the raw material required by poultry and have enumerated some of the more important foods which are available to the farmer in preparing rations for his flock. A few cautions regarding the selection of materials will not be out of place.

1. **Select feeds that the fowls relish.** Rye is not relished by poultry and should not be fed, if other grains are available. Fed alone it will poison the flock and cause many losses. Oats and buckwheat are not relished on account of their indigestible hulls. Therefore they should be provided in inviting form. Palatability is the first requisite.

2. **Select feeds that are easily digested.** Millet seed is palatable but not easily digested, so is of little value as a poultry feed. A food that has an excess of crude fiber is difficult of digestion and should be avoided.

3. **Select feeds that are high in nutrient value.** Polished rice is palatable and easily digested but it lacks in protein and is not a safe feed for poultry.

4. **Select feeds that are farm produced as far as possible.** They are more available and less expensive. The concentrates of animal and vegetable origin must be purchased, but why purchase milo maize when corn is at hand.

5. **Select feeds that are free from mold and decay.** This is the path of safety. Moldy and rotten feeds are dangerous and account for a large percent of poultry losses. Wheat must be free from must; corn should be hand-selected and shelled especially for the flock; oats should be examined for musty and rotten kernels. So also all other feeds should be given the closest scrutiny.

How Food is Used

The changes which take place in the raw material, or food, as it is being transformed into the component parts of the body, involve several processes: **Deglutition,** swallowing; **mastication,** pulverizing the food; **digestion** or dissolving and chemically changing it so that it can be transfused through the walls of the blood vessels; **absorption,** taking it up into the blood and lymph; **circulation,** transferring it to the parts of the body where it is needed for repair and growth; **assimilation,** converting it into the substances which make up the organism; and **oxidation,** a process by which cell substances and organic compounds in the blood are united with oxygen, producing heat and energy; and **excretion,** by which the waste matter is thrown off from the body.

These processes have been studied to some extent in Chapter II. Our discussion here pertains more especially to digestion, assimilation and oxidation.

Another Practical Lesson

It has been pointed out that the albumin and other proteids of the food are changed into peptone by the gastric juice and this is carried by the portal circulation to the liver where it is changed back to albumin. The gastric juice secreted by the stomach and gizzard contains three digestive principles—hydrochloric acid, pepsin and rennin. The rennin coagulates the albumin and the pepsin changes it to peptone, but this process cannot be carried on without the aid of the hydrochloric acid.

Now the hydrochloric acid is manufactured from the salt which is found in the blood. This suggests the importance of feeding salt in the daily ration. Its aid to digestion is here indicated.

Another digestive process is the change in the carbohydrates (starches and sugars) to glucose. This is also transported to the liver and there converted into glycogen, which is taken up by the blood.

Still another change in the food is accomplished by the pancreatic and intestinal juices by which the fats are saponified and emulsified so that they can be taken into the lymphatic circulation and by this emptied into the general circulation.

What Becomes of Digested Food?

The **emulsified fat** is oxidized in the lungs, producing heat and energy. It is probably all oxidized, as very little is found in the blood after it leaves the lungs. It is not used to make fatty tissue. That is made from the other elements of the food in the body itself.

The **glycogen,** representing the carbohydrate of the food, is oxidized in the blood. It is the great source of heat and energy. Not all is oxidized there, but a portion of it goes to the cells and is used to manufacture fat.

The **albumin** of the blood is carried to the cells throughout the body and used to repair waste and build new cells and tissues for the body. Some of it is oxidized, but its chief function is that of supplying material for growth and maintenance. It is a nitrogen-bearer, and there can be no protoplasm or living matter without nitrogen. It is even used in the manufacture of fat, and this explains why fowls fatten readily when there is a liberal allowance of protein in the food.

Composition of the Animal Body

The elements that enter into the composition of a fowl's body are: Nitrogen, hydrogen, oxygen, carbon, calcium, phosphorus, sodium, potassium, magnesium, iron, chlorine, sulphur, silicon and fluorine. These elements also enter into the composition of an egg. In the body they are combined in many substances, some of them very complex. The process by which these elements and compounds are made a part of the living body is called **assimilation.** **Oxidation,** on the other hand, is a process of burning or tearing down. The chief substances produced in this process are carbonic acid gas and water.

The Kind of Food Required

Evidently a perfect ration should contain all of the above named elements. We are accustomed to emphasize the protein and carbohydrate content of the food but forget that the fowl just as surely needs water and the mineral salts in its ration.

Water is a food because it adds to the weight of the body and is used to manufacture some of the compounds of the body.

The mineral salts, such as salt, carbonate of lime, phosphate of lime, sodium carbonate, sodium phosphate, and magnesium phosphate are truly foods for they enter into structures of the body.

Food Constituents

The substances entering into the composition of foods are classified as **nitrogenous** and **non-nitrogenous.** The nitrogenous substances are known as proteids. As the name implies, they contain nitrogen. The non-nitrogenous substances include the ash and the carbohydrates and the fats. Water enters into the composition of all foods but not in sufficient quantity to meet the demands of the animal body.

Crude fiber is the indigestible portion of the food and is composed chiefly of cellulose.

The following table will help to fix this analysis in the mind.

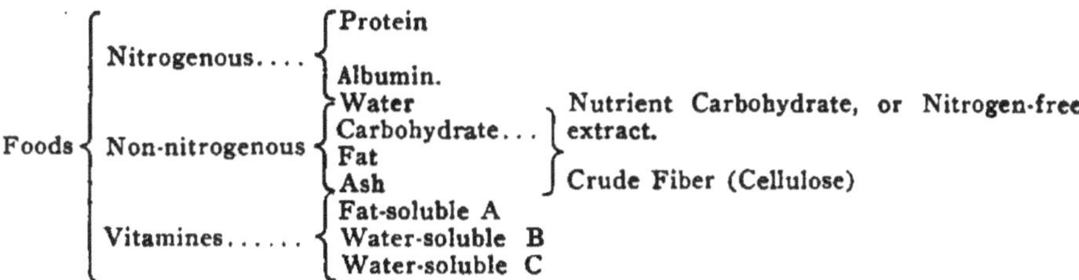

Foods
- Nitrogenous....
  - Protein
  - Albumin.
- Non-nitrogenous
  - Water
  - Carbohydrate... — Nutrient Carbohydrate, or Nitrogen-free extract. / Crude Fiber (Cellulose)
  - Fat
  - Ash
- Vitamines......
  - Fat-soluble A
  - Water-soluble B
  - Water-soluble C

Protein. This is the most important solid constituent of the food. It comprises 20 per cent of the fowl's body. It is necessary to the production of living matter, to cell-multiplication and growth, and to the formation of such tissues as blood, nerve and muscle. It occurs in nearly all grain and animal feeds. Those feeds which contain a large percentage of protein are called proteids. Illustrations are found in tankage, meat scrap, fish scrap, oil meal, cottonseed meal and milk products.

Albumin. This is one of the proteids, but is given special consideration on account of its peculiar properties and its prominence in the tissues of the body, in the blood, and in the composition of the egg.

Water. About 55 per cent of the fowl's body is water. Every cell of its organism cries out for water. Without water an animal soon perishes. It comprises 66 per cent of the composition of an egg.

Egg production ceases when the water supply is cut off. It constitutes 87 per cent of the composition of milk, so that when milk is fed liberally, as well as succulent green feed, the demand for water is decreased. In the body it dissolves the food, aids in absorption, serves as a carrier for the solid principles of the blood, makes the tissues soft and pliable, and enters into some of the chemical changes which are constantly going on.

Carbohydrates. These foods supply heat and energy for the body by oxidation, and the surplus is used in the production of fat. A carbohydrate is composed of carbon, hydrogen and oxygen. The hydrogen and oxygen are always in the proportion found in water, so that when it is oxidized, the oxygen uniting with the carbon to produce carbonic acid gas, water becomes the residue. Carbohydrates abound in all grains and their by-products. In cold weather more carbohydrate is required than in summer. It does not occur in the structures of the body but is found in the blood as glucose, or as glycogen, and in the egg as glucose.

Crude fiber (cellulose) is a carbohydrate, but is insoluble, and, therefore, is indigestible. About 6 per cent of the dry mash may be crude fiber.

This is a convenient type of small colony house

Fat. Fat is composed of carbon, hydrogen and oxygen, but the hydrogen and oxygen are not in the proportion found in water. Fat, taken as food, is oxidized in the lungs to produce heat and energy. It occurs as vegetable oils in grains and seeds and as animal fat in meat scrap, bone meal, etc. The fat in the fowl's body, deposited in the intercellular spaces and as masses of adipose tissue, is manufactured in the body and serves as a reserve supply of fuel for the body and as material for the manufacture of eggs. A hen to lay well should carry a good supply of fat. Lean hens with thin breast bones are invariably poor layers.

Ash. This term comprises the mineral salts and charcoal. They enter into the bones and other structures and form the shell of the egg. Grains and animal feeds usually provide sufficient ash to meet the hen's needs. Some is obtained from the water and some from the soil. Fowls are fond of eating soil, especially if released from confinement, indicating that the ash element is lacking in their food. Charcoal is a corrective, is not digested, but should always be supplied, as it absorbs poisonous gases, aids indigestion, and contributes to the health of the fowl.

Vitamines

Vitamines. A vitamine is a substance whose presence in the food is essential to growth and health. There are three substances of this nature, known as Fat-soluble A., Water-Soluble B, and Water-Soluble C. Hopkins says: "No animal can live on a mixture of pure protein, fat and carbohydrate; and even when the necessary inorganic material is supplied the animal still cannot flourish." Not only is a balanced ration important, but the growth principles must be present in due proportion or the results will be disappointing.

Vitamines have not been chemically analyzed. Their existence is known by experiment, their absence invariably resulting in such serious diseases as rickets, scurvy and polyneuritis (Beri-beri).

Fat-soluble A occurs in such feeds as whole milk, eggs, whole grains, linseed, cabbage, lettuce, carrots, and potatoes, and meats. Its absence results in rickets, a disease affecting the whole body.

Water-soluble B is found in skim-milk, eggs, whole grains, bran, linseed, and most vegetables. Its absence results in Beri-beri, or polyneuritis. A fowl fed solely on polished rice contracts this disease and may be cured by feeding whole rice or any of the feeds named above.

Water-soluble C is found in milk, cabbage, turnips, carrots, potatoes, beets, lettuce and fruits. Feeding substances deficient in this vitamine results in scurvy, and the cure of this disease is secured by correcting the ration. Water-soluble C is found in all fruit juices and more or less in nearly all vegetables and fresh meat.

All of the above nutrients are important. Even the crude fiber is of value in limited quantity, as it furnishes material to stimulate the peristaltic action of the intestines. In excess, however, it hinders digestion. The enumeration of poultry feeds given above comprises chiefly those produced on the farm. By careful planning the poultry keeper can produce his own feeds, thus reducing expenses and enabling him to prepare his own rations. Grain mixtures and mashes prepared at home are not only of known composition, but the quality of the nutrients can be known and regulated.

Chapter IV

Feeding

PROPER feeding is the chief secret of success in poultry raising. Large losses of young chickens, failure to get winter eggs, and many diseases and other troubles are due to improper feeding.

We must first understand the needs of the fowl's organism and then study faithfully how to supply those needs. Some of the rations which follow have been worked out in detail so that the reader can understand how the nutritive ratios are determined. All the rations, except the one pertaining to fattening, require a double mixture, i.e., a grain mixture and a mash mixture.

It would be very difficult to compound a balanced ration such as the hen requires for egg-production from whole grains because they would be deficient in protein, but when we can add to the grain mixtures such protein concentrates as are found in animal and vegetable meals, it is quite easy to provide a balanced ration for the laying hen.

The use of such a ration explains why the hens lay in winter. Under the old system of feeding whole grains, eggs were a great rarity in winter. Now it is the rule for the farm flock to give a good account in the months when eggs are supposed to be scarce. Hens always lay in summer time because then they can secure insects and green feed and thus balance the grain ration that the farmer provides. Under scientific feeding they will respond just as faithfully in the winter season.

Principles of Feeding

If you will note the following outline of the composition of the body and egg of a fowl you can understand that scientific feeding consists in bringing to the flock all the elements needed for growth and maintenance and production, and these elements must be provided in a manner economical and conducive to health.

	Water	Protein	Fat	Ash
Fowl, per cent	55	20	19	6
Egg, per cent	66	13	9	12

Body growth and maintenance come first. Unless there is a surplus of material above that required for heat, energy, growth and maintenance there can be no production of eggs. In laying down the following principles of feeding the demands of the fowl have been consulted.

Balanced Rations

A **balanced ration** is necessary, that is, a ration that supplies the food elements that are needed, and in the proper proportion. If poultry raisers would feed only those rations that meet the actual needs of the fowls in each season, their troubles would cease. The medicine chest would be forgotten and the question, "Does poultry pay?" would receive an immediate answer in the affirmative. This is the secret they long have sought. Detailed grain and mash rations for all conditions are given on pages 46 to 51.

Supplementary Feeds

In addition to the regular ration selected, certain supplementary materials should be provided:

Grit is actually needed for grinding the feed. Its presence in the crop is not absolutely needed for the life of the bird, for fowls have been known to thrive for months without it. It has been proven, however, that a supply of grit means more rapid mastication, more complete digestion and greater thrift.

If grit material is made of quartz or granite it is insoluble in the digestive fluids, and a small quantity will last for a long period. A fowl may be deprived of grit for months and yet a quantity will be found in the crop. Oyster shell serves as grit for a brief period only, as it is dissolved by the hydrochloric acid of the digestive fluids. Good sharp grit should always be in reach of the flock.

Water must be classed as a food. As it comprises 55 per cent of the fowl's body and 66 per cent of the composition of an egg, its importance is evident. It is true that a bird can live for a considerable period without water, but it cannot produce eggs without it nor can it live indefinitely. I have known baby chicks to thrive for two weeks without any water outside of what they obtained in their daily ration of food.

You consider this a cruel experiment, but probably they did not suffer seriously, as all feeds contain a large percentage of water and this is constantly being set free in assimilation and taken into the blood. The importance of clean water in clean vessels cannot be emphasized too strongly. Water gives plumpness to the body, aids digestion, takes part in the processes of assimilation, is the carrier of waste matter to the lungs and kidneys, and, therefore, is essential to a healthy body.

The best method to supply water is in open vessels. These should be placed on elevated platforms in such a position that the fowls cannot get into them with their feet. The vessel should have sloping sides so that in case of freezing it will not be easily broken and the ice can be easily removed. Such a vessel is quickly cleaned and on a platform as suggested does not become foul from the litter.

Milk is of great value for growing stock. It can be fed sour or sweet, but it is advised not to change from one kind to the other. For all seasons sweet milk has preference. Sweet milk can be used as soon as separated and, if fed in the morning, it will usually keep sweet until all is consumed. It is of great value in preparing moist mashes for growing stock and fattening rations. It contains considerable protein but this is balanced by the carbohydrate of the succulent feeds. Chicks fed on milk grow rapidly and are thrifty. It is claimed that the lactic acid of milk holds in check the bacteria of white diarrhea. It is certainly true that if a brood of chicks once becomes infected with bacillary or coccidial diarrhea it has little value as a cure.

Ash comprises the mineral salts. Of these phosphate of lime and carbonate of lime are of especial importance in forming bone and the shell of eggs. A good source of phosphate of lime, or phosphoric acid, is bone meal or granulated bone. It will pay to keep this before growing chicks constantly. It contains 45 per cent of phosphoric acid and in a form that is easy of assimilation. It also contains protein, which adds to its value as a food. The lime for the egg shell is obtained by feeding oyster shell. If this is withheld it means thin shells and fewer eggs.

Green feed of a succulent nature is essential to the health of a fowl. It is valuable not only because it contains water and other nutrient substances but it contains the vitamines, or growth principles so necessary to vitality and growth. The white potato is anti-scorbutic and will ward off the disease known as scurvy. Beets, carrots, Swiss chard, dandelion leaves and rape will help prevent sore eyes due to dietary deficiency; mustard leaves have a tonic effect and make a good green food.

Quality of Feed

Sound and clean grains should be selected for the ration. Sometimes shriveled and chaffy grain, caused by weather conditions, contains a larger protein content in proportion to weight than heavier grain and is perfectly safe for feeding, if free from disease. Mill feeds purchased for the dry mash should be examined with the closest scrutiny. The odor and texture will be a guide as to quality. Musty, moldy and spoiled feeds should be rejected. Rotten potatoes will cause ptomaine poisoning. Rotten and moldy oats will cause Aspergillosis. Tainted meat scraps have caused untold losses. Putrid milk sometimes brings on an epidemic of disease. Many diseases are communicated through foul water.

Crude fiber should be avoided. It is largely cellulose and indigestible.

How Much Feed and When?

Overfeeding results in serious disturbances of the digestive system. An excess of protein means liver and kidney disease. An excess of any one kind of feed, such as corn, often leads to serious digestive disorders. Underfeeding is as serious as overfeeding, for it means a stunted growth and weakness that render the fowl suceptible to disease.

The quantity must be determined by the attendant. He must study the needs of the flock and feed only what will be consumed without waste. The average daily ration of 100 Leghorn hens is 15 pounds of grain mixture and mash. This would mean 10 pounds of grain and 5 pounds of mash. The average daily amount consumed by 100 hens of the dual purpose type is about 18 pounds. This would mean about 12 pounds of grain mixture and 6 pounds of mash. On free range a Leghorn hen will eat about 55 pounds of feed in a year and a dual purpose hen about 75 pounds.

There is no fast rule as to the quantity of feed to be given. A heavy layer will eat much more than a poor layer. Consumption will be greater in extremely cold weather than in mild weather. Consumption will be greater in the spring when the whole flock is busy in egg-production than in the fall when few are laying. Here is where the personal equation enters. and the attendant must win or lose according to the judgment he uses.

Cleanliness

This is one of the cardinal principles of good feeding. A clean feeding place is vital. Damp and moldy litter is dangerous. A feeding floor covered with the dropping of diseased chicks means speedy ruin to the whole brood. When the oppressive days of summer come it is no pleasant task to clean out the brooder or the colony house and replenish the litter, but it must be done promptly and regularly or failure will follow. Watering vessels can be kept clean by placing them on elevated platforms. The same is true of hoppers and self feeders for grit, charcoal and mashes.

Exercise and Feeding

As far as possible feed should be given so as to encourage exercise. Even the mash can be placed so that the fowl must use some physicial exercise to get it.

The grain mixture should be fed in moderately deep litter six to eight inches in depth. Care should be used in selecting the litter. For young chicks short-cut alfalfa, alsike clover, or short-cut rye straw will be found safe. Wheat and oat straw often bear the spores of mold and smut and rust. which frequently produce fatal disease.

Rye straw is usually bright and clean and if run through the silo cutter will make a very fine litter. Chaff gathered around the threshing machine is usually dusty, and is unsafe to use. If the floor of the scratch pen is inclined to be damp the litter should be shallow to allow the dampness to dry out. The fowls in the pen with damp litter are the ones to get out of condition. The litter should be changed monthly while the flock is in winter quarters.

How to Feed

There are about as many systems of feeding as there are poultry keepers. Every farmer's wife has her own system, and she generally wins average success. The systems given below are not perfect, but if you give them a fair trial you will not be disappointed.

Feeding the Baby Chick

Feed nothing until the third day after the chick is hatched. Just before the chick is ready to break the shell the yolk of the egg is absorbed into the abdomen. This is a provision of nature to furnish nourishment for the chick during the first few days of its growing life. During these few days the yolk is absorbed into the circulation and assimilated.

If the chick is fed before that process is completed, which requires about 72 hours, the process of absorption is checked and the yolk remains in the abdomen, a menace to its health and growth. Many chicks that perish, if examined, would be found to contain the unabsorbed yolk. At the end of this period, or at the close of the third day, give a light feed of rolled oats and give sweet milk for drink.

The feed should be very simple for the first two days, nothing but rolled oats, with milk in the forenoon and water in the afternoon. If the plan of removing the milk at noon, cleansing the vessels, and replacing with water in the afternoon is followed throughout the feeding period, there will be little danger of harm from putrid milk. We advise sweet milk because it is just as valuable as sour milk and is available at all seasons.

After the second day of feeding, place the chicks on Ration No. 1, found on page 46. Rolled oats or pinhead oats constitute the scratch ration, and should be thrown in shallow litter to induce exercise. A feeding box about three inches deep and three feet square would answer well for 100 chicks. This could be removed, cleaned and supplied with fresh litter as required. The mash portion of the ration should be placed in a hopper upon an elevated platform so as to keep it clean. Self feeders for this purpose can be purchased at trifling cost. Near the mash feeder should be placed a hopper with three compartments containing grit, charcoal, and granulated bone.

At the end of the second week change gradually to Ration No. II (page 46). This means a change from rolled or pinhead oats to whole wheat as the scratch feed, and the change can be made by adding a little wheat to the oats, then increasing the wheat until the oats can be omitted.

At the end of four weeks change to Ration No. III (page 46). Note that the mash mixture for No. III is the same as for No. I and No. II, so that no change will be required in that portion of the ration. The grain mixture, however, now consists of corn, wheat and oats, a less expensive ration and one presenting a greater variety.

After the first week green feed should be supplied. Sliced raw potatoes will be greatly relished by the chicks. The tops of sprouted oats or rape also serve well. Dandelion leaves are especially recommended on account of their favorable action on the liver. Beet pulp is a good succulent food. Lettuce and rape are recommended by some poultry keepers. Swiss chard is worth considering.

If Ration No. IV is used the same general system of feeding is recommended.

How to Feed for Egg Production

Use Rations No. V or VI (page 49) for winter feeding. The grain mixture should be thrown in deep clean litter to promote exercise. The mash should be kept before the fowls continually in hoppers on elevated platform. Otherwise the hoppers will be filled with litter. Near the mash should be a four-compartment hopper with grit, charcoal, oyster shell, and granulated bone. For litter there is nothing better than bright alsike hay. If this cannot be obtained, chopped rye straw or bright wheat straw is advised. If this cannot be obtained oat straw should be used as a last extremity.

The aim should be to secure litter that is not dusty or moldy. Change the litter once a month during the winter and once in three months when the fowls are on free range. For green feed mangel wurzels once a day are valuable.

Sprouted oats or sprouted rye are greedily consumed by the fowls. In sprouting oats, mold often forms. This can be avoided by washing the trays with a solution of formaldehyde and by adding to the water in which the oats are to be soaked over night a few drops of formaldehyde. Cabbage, carrots, turnips, pumpkins, alfalfa leaves and alsike leaves all give good results as succulent feeds.

When the days grow warmer a change should be made to Ration No. VIII (page 50). This is a narrower ration but is adapted to summer feeding. Keep the self-feeder full of mash and feed the grain mixture once a day.

Feeding the Breeding Stock

It is not customary to make any distinction between the **utility** layers and the breeders in the matter of feeding, but where eggs are to be sold for hatching, or the poultry keeper wishes to raise a large supply of breeding stock, the care and feeding of the foundation stock are certainly important.

Where stock is to be used solely for the production of market eggs, forced feeding should be used for the first year or two, and then they should be fattened for market. In the case of breeding stock, however, where vigorous offspring are desired too much forcing may bring disaster. Whole grains are indicated, and the mash should not be too rich in protein. Ration No. VII is recommended.

This is a wide ration and will probably bring the breeding stock to laying condition about the 15th of February. This is about the time when the farmer begins to think of filling the incubator. His flock has not been weakened by forced feeding for eggs, but is in the pink of condition.

How to Feed During the Molt

For the formation of feathers a ration rich in nitrogen is required. Ration No. IX (page 51) is advised. It should follow the summer ration, beginning about the first of September and continuing till the last of November. If the fowls are confined, the usual allowance of succulent feed and ash should be provided.

Feeding for the Market

Capons, surplus cockerels and culls from the laying flock should not be shipped to market without conditioning. Even confinement in a room with whole corn, water and grit would be better than no conditioning at all. If Ration No. X (page 51) is used quick results will be obtained.

The crate method of feeding, with slatted bottoms to insure cleanliness, and a feeding trough outside the crate, is a common method of feeding. A small room with litter and an elevated platform for the feeding trough and other vessels will answer as well. The mash should be given three times a day, all that the fowls will consume in twenty minutes. As soon as the fowls are fat, market promptly. In crate fattening Leghorns, provision should be made for exercise or the results will be disappointing.

Constructing a Ration

Balanced rations are determined by the ratio existing between the protein content of the ration and the starch, sugar and fat (nutrient carbohydrate). This ratio is called the **nutritive ratio.**

For example, a balanced ration for growing chicks requires that the starch, sugar and fat in the combination of feeds

Dinner time in the poultry yard

should be four and one-half times greater in weight than the protein. The nutritive ratio therefore is 1 : 4½.

To illustrate further, in a certain combination of feeds which furnish a balanced ration for growing chicks it is found that there are 10 pounds of protein and 45 pounds of starch, sugar and fat. The ratio between the two is therefore 10 : 45. Reducing this by dividing the ratio by 10, we get a nutritive ratio of 1 : 4.5.

To illustrate further, in a certain combination of feeds ratio should be approximately 1 : 5. It may be a little more or a little less. If the nutrients are supplied in that ratio the hen will see that her ration is balanced by selecting the feeds she needs. She cannot be fed with a spoon, or by exact rule or measure. . If the nutrients are placed before her in approximately the right proportion her own instincts will guide her in selecting the food required.

To furnish a ration suited to fattening fowls for market, it has been found that the nutritive ratio should be 1 : 3, i.e., there must be a smaller proportion of starch, sugar and fat than in the nutritive ratio for the laying hen.

How is the Nutritive Ratio Determined?

From the observations already made it is easy to understand that the nutritive ratio is the relation existing between the protein and the carbohydrate, it is the comparison between the weight of the nutrient protein and the nutrient carbohydrate.

How is this ratio determined? Simply by determining the total weight of protein in all the materials of the ration and then the weight of the carbohydrate in the same materials. When that is done the comparison is easily made by dividing the ratio by the number representing the weight of the protein. To determine the weights in question it is necessary to refer to the table showing the percentages of nutrients in each variety of food.

In determining the carbohydrate it is necessary to reduce the fat to terms of carbohydrate. This is done by multiplying by 2¼ and adding the product to the weight of the nitrogen-free extract. The reason for this is that fat has the power to produce two and one-fourth times as much heat and energy as the same weight of nutrient carbohydrate.

Illustration: A ration consists of 10 pounds of corn, 10 pounds of wheat, and 10 pounds of oats. What is the nutritive ratio?

By referring to the table given below, showing the percentages of nutrients, it is possible to determine the weight of each nutrient. The following illustration shows how the problem is worked out and the nutritive ratio determined.

Grains	Protein	Nutrient Carbohydrate	Fat
10 lbs. Corn	1.05 lbs.	6.96 lbs.	.54 lbs.
10 lbs. Wheat	1.19 lbs.	7.19 lbs.	.21 lbs.
10 lbs. Oats	1.18 lbs.	5.97 lbs.	.50 lbs.
Totals	3.42 lbs.	20.12 lbs.	1.25 lbs.
Reducing fat to Carbohy.		2.81	
Adding	3.42	22.93	
Dividing by 3.42=	1	6.7 =the nutritive ratio.	

To convert the 1.25 lbs. of fat to terms of carbohydrate we multiply by 2¼. This gives 2.8 lbs. which is added to the 20.12 lbs. of nutrient carbohydrate (often called nitrogen-free extract) and this gives us 22.93 lbs., which represents the carbohydrate in the 30 lbs. of food. Our ratio therefore is 3.42 : 22.93. Dividing this ratio by 3.42 we get the nutritive ratio, which is 1 : 6.3.

Wide and Narrow Nutritive Ratios

It will be observed in the ration just given that the amount of carbohydrate is much larger than the amount of protein, much larger than in the nutritive ratio for laying hens, which is 1 : 5. Such a ration is said to have a wide nutritive ratio. On the other hand the nutritive ratio for growing chicks has a smaller amount of nutrient carbohydrate than found in the nutritive ratio for the laying hen, and such a ratio is said to be narrow.

A good day's work

Table I. Composition of Poultry Feeds

Feeds	Water Percent	Ash Percent	Protein Percent	Crude Fiber Percent	Nutrient Carbohydrate Percent	Fat Percent
Whole grains:						
Corn	10.9	1.5	10.5	2.1	69.6	5.4
Kafir Corn	12.8	2.1	9.1	2.6	69.8	3.6
Barley	10.9	2.4	12.4	2.7	69.8	1.8
Oats	11.0	3.0	11.8	9.5	59.7	5.0
Wheat	10.5	1.8	11.9	1.8	71.9	2.1
Buckwheat	12.6	2.0	10.0	8.7	64.5	2.2
Sunflower seed	8.6	2.6	15.3	29.9	21.4	21.2
Soy bean	8.7	5.4	36.3	3.9	27.7	18.0
Ground grains:						
Corn meal	15.0	1.4	9.2	1.9	68.7	3.8
Barley meal	11.9	2.6	10.5	6.5	66.3	2.3
Soy-bean meal	10.2	5.0	35.9	3.4	28.0	17.5
Gluten meal	8.6	.6	30.0	2.6	49.2	8.8
Gluten feed	8.1	1.3	23.2	6.4	54.7	6.3
Wheat bran	11.9	5.8	15.4	9.0	53.9	4.0
Wheat middlings	12.1	3.3	15.6	4.6	60.4	4.0
O. P. Linseed meal	9.2	5.7	32.9	8.9	35.4	7.7
Foods of animal origin:						
Meat scrap	7.9	17.4	49.7	....	....	18.5
Meat meal	6.3	....	48.4	....	....	12.9
Blood meal	10.6	4.6	75.7	1.3	1.4	7.1
Tankage	10.0	10.0	60.0	3.0	....	0.5
Fish scrap	7.5	6.0	42.0	....	....	17.0
Whole milk	87.2	.6	3.6	....	4.9	3.7
Skim milk	90.6	.7	3.1	....	5.3	.3
Buttermilk	90.1	.7	4.0	....	4.0	1.2
Granulated milk	28.5	3.6	13.7	....	51.1	3.1
Green feeds:						
Green alfalfa	80.0	1.8	4.9	7.9	....	.1
Alfalfa meal	11.9	7.1	14.1	27.1	37.3	2.4
Green clover	70.8	2.1	4.4	8.1	13.5	1.1
Clover meal	10.0	8.1	16.3	17.8	46.0	1.7
Potatoes	78.9	1.0	2.1	.6	17.3	0.1
Mangel beets	90.9	1.1	1.4	.9	5.5	0.2
Dry beet pulp	8.0	5.4	9.5	15.4	61.3	0.4
Onions	87.6	0.6	1.4	0.7	9.4	0.3
Turnips	90.5	0.8	1.1	1.2	6.2	0.2
Carrots	88.6	1.0	1.1	1.3	7.6	0.4
Cabbage	90.5	1.4	3.8	1.5	2.4	0.4
Lettuce	95.5	0.8	1.6	0.5	1.0	0.2
Swiss chard	87.8	2.4	4.4	2.9	2.5	0.4

The above percentages are in the main supplied by the U. S. Department of Agriculture. As food materials vary in composition, owing to conditions over which the analyst has no control, no two analyses of the same substance will be exactly the same. Those given above are sufficiently accurate for all practical purposes.

Using the table given, the poultry keeper can compound his own rations. Knowing the object he wishes to attain, whether to grow the chick or provide for the laying hen or to fatten for market and knowing the nutritive ratio required, he can combine a ration from the feeds at hand that will give as good results as by the purchase of more expensive feeds.

Some elements of the ration may have to be purchased, such as protein concentrates and a few vegetable meals, but in the main the grains and green feeds can be produced on the farm. The balanced rations given below have been tried out and we believe will be found safe and reliable.

Rations for Growing Chicks

Ration No. I. For first two weeks. Nutritive ratio required 1 : 4.5.

Scratch feed:	Lbs.	Protein	Nutrient Carbohydrate	Fat
Rolled or pin-head oats	100	15.00	66.00	8.00
Mash:				
Corn meal	20	1.84	13.74	0.76
Oat meal	20	2.94	13.48	1.42
Bran	20	3.08	10.78	0.80
Wheat middlings	20	3.12	12.08	0.80
Meat scrap	20	9.94	0.00	3.70
Totals	200	35.92	116.08	15.48
15.48 lbs. Fat × 2¼ =			34.83	
Adding		35.92	150.91	
Dividing by 32.92 =		1 :	4.5, the nutritive ratio.	

Ration No. II. For third and fourth weeks. Nutritive ratio required 1 : 4.5.

Scratch feed:	Lbs.	Protein	Nutrient Carbohydrate	Fat
Wheat	100	11.90	71.90	2.10
Dry Mash:				
Corn meal	20	1.84	13.74	0.76
Oat meal	20	2.94	13.48	1.42
Wheat bran	20	3.08	10.78	0.80
Wheat middlings	20	3.12	12.08	0.80
Meat scrap	20	9.94	0.00	3.70
Totals	200	32.82	121.98	9.58
9.58 lbs. fat × 2¼ =			21.55	
Adding		32.82	143.53	
Dividing by 32.82 =		1 :	4.4, the nutritive ratio.	

Ration No. III. After fourth week.

Scratch feed:	Lbs.	Protein	Nutrient Carbohydrate	Fat
Whole wheat	40	4.76	28.76	0.84
Cracked corn	40	4.20	27.84	2.16
Hulled oats	20	2.94	13.48	1.42
Dry Mash:				
Corn meal	20	1.84	13.74	0.76
Oat meal	20	2.94	13.48	1.42
Bran	20	3.08	10.78	0.80
Wheat middlings	20	3.12	12.08	0.80
Meat scrap	20	9.94	0.00	3.70
Totals	200	32.82	120.16	11.90
11.09 lbs. Fat × 2¼ =			26.77	
Adding		32.82	146.93	
Dividing by 32.82 =		1 :	4.5, the nutritive ratio.	

Rations I, II, and III are designed to be used in the order suggested. Note that the same mash mixture answers for all three rations. By following the system here given the chicks are provided with nourishing food, make rapid growth, and are early brought to a ration of whole grains. If milk is given it may be sweet skim milk, which is the most convenient for all seasons. It should be placed before the chicks in the morning and removed at noon, to be followed by water in the afternoon. Succulent green food should be supplied after the second day. A hopper containing chick size charcoal, chick grit and granulated bone should be accessible at all times. The granulated bone will furnish the ash needed for the development of bone and other tissues.

Fine charcoal should be added to all the mashes, about six pounds to 100 pounds of mash. No grit should be given except in the hopper, but in ration No. III one-half pound of salt may be added to each 100 pounds of mash. The salt must be fine and free from lumps. These rations have been developed on the understanding that chicks will eat equal quantities of scratch feed and mash. Their own instincts and appetites will help them balance their ration if given the opportunity. Succulent feed should not be neglected, sliced potatoes, beets, dandelion leaves, tops of sprouted oats, lettuce, mustard or swiss chard.

Ration No. IV. From baby chick to maturity. Nutritive ratio 1 : 4.5.

	Lbs.	Protein	Nutrient Carbohydrate	Fat
Scratch feed:				
Cracked wheat	40	4.76	28.76	0.84
Fine cracked corn	40	4.20	27.84	2.16
Pin-head oats	20	2.94	13.48	1.42
Mash:				
Bran	45	6.93	24.25	1.80
Oat meal	40	5.88	26.96	2.84
Meat scrap	15	7.45	0.00	2.77
Totals	200	32.16	121.29	11.83
11.83 lbs. Fat =			26.62	
Adding		32.16	147.91	
Dividing by 32.16 =		1 :	4.6 , the nutritive ratio.	

The above ration can be used from the first day, but it would be well to start the chicks on rolled oats or bread and hard boiled egg for the first two days. The directions given above regarding milk, charcoal, granulated bone, grit and green feed should not be overlooked. If the chicks have free range, the supply of green feed can be limited. Green feed will overcome scorbutus (scurvy) and lameness and help prevent sore eyes due to dietary deficiency.

Rations for Egg Production

If hens are to be forced for egg production a narrower ration may be used than when they are to be fed for breeding purposes. If a breeder is fed a forcing ration her vitality may be so reduced that when the breeding season comes her eggs will not be fertile or the chicks that hatch from them may lack in vigor.

It would be wisdom, therefore, to pen the hens intended for breeding purposes in separate pens and allow them a wider ration. The laying hen requires a nutritive ratio of 1 : 5. It has been found in the egg-laying contests that one pound of carbohydrate will produce 3⅓ yolks and that one pound of protein will produce 16⅔ whites. This results after the needs of the hen's body are supplied.

To produce one hundred yolks, therefore, would require 30 pounds of carbohydrate; and to produce one hundred whites would require six pounds of protein. This gives a ratio of one to five, that is, when there is five times as much carbohydrate in the food as protein there will be produced an equal number of whites and yolks. This shows the necessity of a balanced ration, for if the protein is deficient there will be too few whites produced for the number of yolks, and the process of completing the egg will be delayed. The two rations given below will be found helpful. It is generally supposed that a hen will consume the same weight of grain ration as mash. In actual experience, however, a hen usually consumes twice as much grain as mash. This would unbalance our ration and supply the hen with a greater proportion of carbohydrate than the nutritive ratio requires. That these statements are true is attested by the fact that during the winter months, when forcing rations are used, the hen needs more carbohydrate to maintain heat and energy. If left to her own instincts and the materials are placed before her, she will see that the proper balance is maintained for egg production. This cannot be done if the ration is lacking in any essential.

Ration No. V. For winter egg production. Nutritive ratio required 1:5.

	Lbs.	Protein	Nutrient Carbohydrate	Fat
Grain mixture:				
Cracked corn	60	6.30	41.76	3.24
Wheat	40	4.76	28.76	0.84
Totals	100	11.06	70.52	4.08
Fat to carbohydrate			9.18	
Adding		11.06	79.70	
Mash:				
Bran	30	4.62	16.17	1.20
Oat meal	30	4.41	20.22	2.13
Alfalfa meal	20	2.82	7.46	0.48
Meat scrap	20	9.94	0.00	3.70
Salt, ½ lb.		....	....	....
Totals	100	21.79	43.85	7.51
Fat to carbohydrate			16.90	
Adding		21.79	60.75	
Dividing by 2 =		10.89	30.37	
Adding grain mixture		11.06	79.70	
Totals consumed =		21.95	110.07	
Dividing by 21.95 =		1 :	5, the nutritive ratio.	

Note.—The nutritive ratio of the above ration, based on the equal weights of grain mixture and mash is 1 : 4.3. As in actual practice the hen consumes only half as much mash as grain mixture, we must divide the protein and carbohydrate of the mash by two and add the quotients to the corresponding weights in the grain mixture. The resulting weights will enable us to obtain the nutritive ratio of the actual nutrients consumed by the hen.

Ground oats can be used instead of oat meal, but add 9 per cent to allow for crude fiber.

Ration No. VI. For winter egg production. Ratio required 1: 5.

	Lbs.	Protein	Nutrient Carbohydrate	Fat
Grain mixture:				
Cracked corn	75	7.87	52.20	4.05
Whole wheat	50	5.95	35.95	1.05
Hulled oats	25	3.67	16.85	1.77
Totals	150	17.49	105.00	6.87
Fat to carbohydrate			15.46	
Adding		17.49	120.46	
Mash:				
Wheat bran	25	3.85	13.47	1.00
Wheat middlings	25	3.90	15.10	1.00
Corn meal	25	2.30	17.17	0.95
Oat meal	25	3.67	16.85	1.77
Mealed alfalfa	20	2.82	7.46	0.48
Meat scrap	25	12.42	0.00	4.63
O. P. Oil meal	5	1.64	1.77	0.39
Salt, ½ lb.		....	....	....
Totals	150	30.60	71.82	10.02
Fat to carbohydrate			22.99	
Adding		30.60	94.81	
Dividing by 2		15.30	47.40	
Adding grain mixture		17.49	120.46	
Totals consumed		32.79	167.86	
Dividing by 32.79		1 :	5, the nutritive ratio.	

Note. This ration is more complex than the preceding but has the advantage in providing greater variety. Charcoal should be added to all dry mashes, about three to six pounds to 100 pounds of mash.

If ground oats or whole oats are used add 9 per cent on account of crude fiber.

Ration No. VII. For summer egg production on free range.

Grain mixture:		Mash:	
Cracked corn	30 lbs.	Bran	50 lbs.
Wheat	40 lbs.	Ground oats	40 lbs.
Clipped oats	30 lbs.	Meat scrap	10 lbs.
	100 lbs.		100 lbs.

This ration gives a nutritive ratio of 1 : 5. Hens on the farm do not always produce well during the summer months. This is often because the ration is neglected. By providing a mash as indicated in self feeders in dry and convenient places and keeping the hoppers full, the hens will give a good account• of themselves. On free range the grain mixture should be given once a day in the evening.

Ration for Breeding Stock

For stock intended for breeding purposes the ration should not be too narrow, as a ration too rich in protein is too forcing. A nutritive ratio of 1:6 is advised during the winter months. Whole grains are best for the breeders, and they should be fed so as to encourage exercise. In the breeding season, when an abundance of eggs is required, the ration can be narrowed to 1 : 5. The following ration is recommended for the breeding stock if they are kept separate from the general flock.

Ration No. VIII. Winter ration for breeding stock.

Grain mixture:		Mash:	
Wheat	35 lbs.	Wheat bran	25 lbs.
Cracked corn	40 lbs.	Ground oats	50 lbs.
Hulled oats	25 lbs.	Corn meal	20 lbs.
		Meat scrap	5 lbs.
	100 lbs.		100 lbs.

This ration has a nutritive ratio of 1 : 6. One-half pound of fine table salt and three to five pounds of charcoal are advised for each hundred pounds of mash.

Ration for the Molting Season

Feathers are rich in nitrogen and sulphur. For their production a narrow ration, or one rich in protein, is required. The molting season usually begins in earnest in September and as that is the season when there are weather changes and the demand of the body is for an increased supply of heat, corn is indicated as a liberal portion of the diet. Probably the best nutritive ratio for the production of feathers is 1:4 or even narrower. The following will be found of value in this critical season.

Ration No. IX. For the molt. Nutritive ratio required, 1:4.

Grain mixture:		Mash:	
Cracked corn	40 lbs.	Corn meal	40 lbs.
Wheat	40 lbs.	Wheat bran	30 lbs.
Hulled oats	20 lbs.	Meat meal	20 lbs.
		O. P. oil meal	10 lbs.
	100 lbs.		100 lbs.

The nutritive ratio of this ration is 1:4.2. Wheat can be used solely, replacing the other grains, and the same ratio will be preserved.

Ration to Fatten Market Fowls

A large percentage of farm poultry goes to market without any preparation. What the farmer loses by this failure to condition his fowls is gained by the packers and others who make a business of fattening the thin stuff which reaches the commission merchant. With little pains and expense the farmer could reap this profit. Fowls that come from the range are not tender, their meat is tough and stringy. A few days in the fattening room or crate makes a wonderful change in the quality of the meat. A fattening ration requires a narrow ratio. In a former chapter it was pointed out that the excess of carbohydrate, over and above the requirements of the fowl's body, is converted into fat; the same is true of the excess of protein. The fattening ration should be rich in carbohydrate, but the protein constituent should not be overlooked. The ratio usually advised is 1:3. Probably a ratio of 1:4 would answer just as well. A very crude method of fattening is to confine the fowls and place before them whole corn and water. The nutritive ratio of corn is 1:7.5. This ration is too wide; more variety is needed. Protein is the great fat former and should be a prominent factor in a fattening ration. The following ration is recommended.

Ration No. X. For fattening for the market.

Corn meal	100 lbs.
Oat meal	100 lbs.
Bran	25 lbs.
Meat meal	50 lbs.
Skim milk or buttermilk	600 lbs.

Sprouted oats are a good source of green feed for winter layers

This mixture provides a nutritive ratio of 1:3. This is very narrow and is, therefore, very rich in protein. If the mash can be cooked before feeding, it will add to its palatability and its digestibility, and thus shorten the period required for fattening. About two pounds of milk should be used for each pound of mash. A small amount of grit should be accessible during the period. The mash should be placed in troughs, just what the fowls will clean up, and supplied about three times during the day. After each feeding any material left over should be removed and the troughs cleaned. This is intensive feeding and will make a severe tax upon the digestive organs. The period required for fattening is about 10 days.

Fowls fattened by this method will be juicy and tender and should command the top price on the market.

Things to Remember

1. A balanced ration is the most economical, insuring rapid growth and profitable production.
2. The regular ration must be supplemented with grit, water, milk, ash and succulent feeds.
3. Dusty, moldy and damp litter is a menace.
4. Cleanliness is a corner stone of successful feeding.
5. Overfeeding is dangerous.
6. Underfeeding leads to disappointment.
7. Regularity in feeding means a contented flock.
8. Home grown feeds are more reliable than others. They have no fillers.

9. Crude fiber is indigestible and should be avoided.

10. The ration should be adapted to the season and the needs of the flock.

Chapter V

The Finished Product

THUS far we have considered the factory with its complicated machinery, the workmen, invisible but countless in number, and the raw material in multiplied forms. Now our study concerns the finished product, consisting of the edible carcass, the egg and the byproducts.

The Carcass

The flesh of poultry is highly prized. A fowl properly conditioned, well cooked and served is both appetizing and nutritious.

The process of fattening has been considered in the preceding chapter. When ready for the market or the table, food should be withheld from fowls for a day until the crop and intestines become empty. They should have an abundance of clean water to flush out the system and at the same time keep the flesh plump and juicy. For the same reason fowls intended for home consumption should be deprived of solid food for at least 12 hours, and provided with plenty of water.

Killing. The method of killing is very simple. Instead of wringing the neck, according to an ancient custom, a process which interferes with free bleeding, a knife is used to sever the arteries in the throat. The blade should be long and sharp. Seize the head in the left hand, and with the right hand insert the blade in the mouth until the point reaches the base of the skull. Make a transverse cut on the left side, severing the arteries of the neck. If there is free bleeding, the blade should then be forced through the slit in the roof of the mouth backwards into the brain and then given a twist. This produces paralysis and means death without pain and with very little struggling. During the process of killing and while being plucked, the fowl should be suspended by a cord. This method of killing results in loosening the feathers so that dry picking becomes an easy task.

Dry plucking is the common rule with turkeys, ducks and geese. If plucking is done promptly after sticking it can be practiced with chickens. It gives the carcass a more attractive appearance. Another method of killing and plucking is to use a table or bench. The beak is fastened to a screw hook. Just under the head of the fowl is a hole in the bench through which the blood passes to a receptacle below. That portion of the bench on which the body rests is padded to prevent bruising. The legs are held with the left hand while with right hand the incision is made on the left side of head just behind the earlobe. Plucking is done immediately. By this method more rapid progress can be made. During plucking the feathers should be sorted and graded. The feet and head should not be severed.

Plumping is accomplished by plunging the carcass into cold water. While the parts are flexible the wings and head should be adjusted so as to make a compact package.

Scalding. Some prefer scalding to dry picking. The water for scalding should be of correct temperature, 180 degrees. If too hot the skin becomes shrunken and is easily torn in plucking, and the carcass becomes discolored. Properly done, however, this method results in a very attractive carcass. Before plumping it is customary to singe the down, and this is best done with a gas or alcohol flame.

Cleaning. All fowls designed for market are left undrawn. They keep better in this condition as they are not exposed to flies and bacteria. In dressing a fowl for home consumption it is customary first to remove the feet and head. Then a transverse slit is made at a point about half way between the lower point of the keel and the vent. It should be about two inches long. From the middle of this incision another cut is made longitudinally to the vent and around the same, so that when the viscera are removed they will come clean. It is a good plan to remove the crop, œsophagus and trachea first. This is done through an incision made on the side of neck. If the crop is removed first it will be much easier to remove the viscera, which will come clean without leakage.

Packing. Dressed poultry can be packed in barrels or boxes. A layer of excelsior or clean straw is laid in bottom of receptacle, then a layer of dressed fowls with feet extending outward, then excelsior, then another layer of carcasses, and so on till the barrel is full. Any empty space at the top should be filled with excelsior, straw, or paper. If any leakage appears, such as might soil other specimens, it is a good plan to plug the vent and mouth with sterilized cotton. After putting on cover and carefully labeling, the package may be shipped by express. A receipt always should be taken. Insist that the expressman weigh the package, so that you will have an official check against the commission man's weights.

Types of Market Poultry

Broilers. A broiler is a young chicken two to three months old and weighing 1½ pounds to three pounds. Broilers, especially those that reach a late spring or early summer market, command high prices and are very profitable. Surplus and undesirable cockerels should be culled and sold when prices are high.

A **squab** is a young pigeon or duck not fully fledged, yet fat and fit for human consumption. Squab farming is practised to considerable extent and profitably in some sections.

A **roaster** is a matured fowl, fat and fit for roasting.

A **capon** is a de-sexed cockerel. Capons are docile and easily handled. They make rapid growth and attain large size, much larger than the standard weight for the breed. The flesh of the capon is superior and it commands the highest market price. The demand is greater than the supply. Cockerels are usually produced at a loss, but when they are converted into capons, on account of the rapid growth, large size, and better price, they become profitable. They should not be marketed until matured and well fattened. In plucking them, the feathers of the upper neck, the flight feathers and the tail feathers should not be removed, the object of this being to distinguish them from other fowl.

Caponizing. Instruments can be secured of any reliable poultry supply house. The time to caponize is from three to six months of age. The operation requires care and some skill, but any person handy with tools can do the work with a little practice.

The bird is placed upon its left side and held to the table by weights attached by cords to feet and wings. The skin over the right side is drawn forward and an incision made between the last two ribs.

The incision should be about an inch long. The spreader is now applied to keep the aperture open. A slit is made in the omentum, or membrane surrounding the intestines, and the intestines are pushed to one side until the testes are seen. One or both of these is seized by the special forceps made for the purpose and by a twist they are severed and then removed. When both glands are removed from the same opening it is advised to remove the lower one first, so that bleeding will not interfere in removing the remaining one. Losses in this operation usually occur by the severing of the spermatic artery which lies behind the glands. In that case the chick bleeds to death, but its carcass is perfectly good for home consumption. As soon as the spreader is removed the skin closes over the opening, and the wound soon heals without any stitching.

Caponizing should be more generally practised. A larger supply would mean an increased demand and greater profits to the industry, so that both producer and consumer would be benefited.

The spreader in placing, tearing open the membranes. Note how the bird is held in position by weights attached to wings and feet

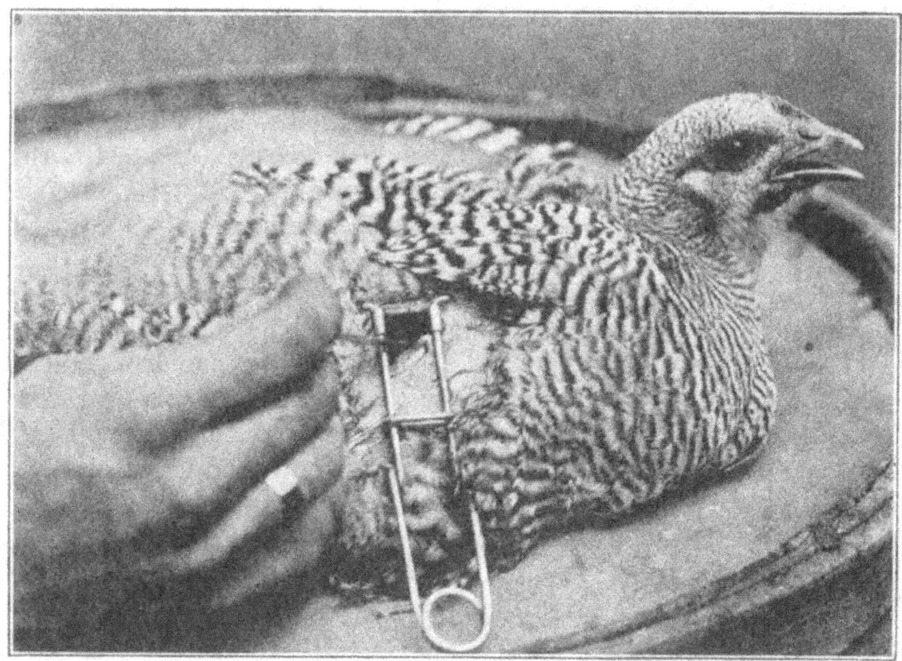

The testicles can be observed between the jaws of the spreader

The Egg

The egg is the chief end of poultry production. It is the real thing in poultry culture, the goal toward which every producer aspires, for it furnishes the promised reward for his labors. It furnishes the ideal food for human consumption, the protein so necessary to build protoplasm, cells and tissues for the human body, and the yolk of the egg is rich in vitamines, or growth principles, without which the animal organism would fail to grow and maintain health.

Many fowl and other farm animals fail to develop properly, becoming emaciated and diseased, because the food supplied is wanting in these growth-producing substances. So it is with children. If not supplied with food rich in growth principles they fail to develop normally.

So also the brain worker or manual laborer requires food of this character that his physical and mental powers may function normally and vigorously. Eggs, therefore, make an ideal food for children and for all upon whom heavy demands are made, either physical or mental. They are not appreciated because they are so cheap. Like milk they provide a perfect food from which all the structures of the animal organism can be produced—bone, muscle, nerve, and connective tissue.

Origin of an Egg

Every animal organism is produced from an egg, that is, from a primordial cell, which corresponds to the initial cell from which every poultry egg is derived. The ovary of a hen contains from 800 to 7000 egg possibilities. An examination in the laying season shows the yolks, or ovules, in various stages of development, from the smallest, which are merely microscopic points, oöcytes, to the fully formed yolk ready to be discharged into the oviduct.

Not all of these undeveloped ovules ripen into mature eggs. The average annual production of a farm hen is not more than 70 eggs. The poultry breeder is content to secure 300 to 400 eggs from each hen in his flock, yet hens have been known to produce more than 1,000 eggs in a lifetime.

The Poultry Department of Purdue University has produced a hen, Joan of Arc, which has laid 1,064 eggs.

The Experiment Station of Oregon Agricultural College has developed twenty hens with trap-nest records of more than 1,000 eggs each. The highest record was made by a hybrid hen containing Barred Rock and White Leghorn blood. In her ninth year she died leaving a total production record of 1,335 eggs. Undoubtedly this is the world's highest egg-record for an individual hen.

How is an Egg Formed?

The **ovary** lies in a delicate membrane known as the **ovi-sac.** The ovi-sac surrounding each developing yolk is generally known as the **follicle.** Under the stimulus of suitable food, exercise and warmth, the initial cells from which the yolks are formed begin to grow. As they develop, successive layers of albumen are deposited. This material is furnished from the blood through the blood vessels of the follicle. Immediately surrounding the yolk a delicate membrane is formed known as the vitelline membrane.

Upon the surface of the yolk lies the **germinal vesicle.** This is the living germ cell with which the sperm cell must fuse to produce the embryonic cell from which the embryonic chick is developed. When the yolk is ripe, or fully matured, the follicle cleaves and allows it to escape. By some strange attraction it is drawn to the mouth of the oviduct, known as the **infundibulum,** and enters the tortuous tube where its development is carried on to completion.

The walls of the oviduct are lined with a network of blood vessels which furnish the material for the further development of the egg. In the upper portion of the oviduct the albumen is secreted and deposited around the yolk in layers. In the middle portion and toward the lower end, the two membranes are formed around the albumen. The outer membrane is covered with carbonate of lime, which is secreted and deposited just before the egg passes into the **cloaca.** This is called the **shell** of the egg.

The cloaca is a pouch at the end of the oviduct sufficiently large to receive the egg. Here a secretion is formed and deposited upon the shell. This film in a measure makes the shell impervious to germs and other harmful substances and prevents the evaporation of the egg contents. The egg is now fully formed and ready to be laid. The time required for the development of the yolk in the ovary is about three weeks; but, after the egg enters the oviduct, only 18 hours are required for its completion. It is evident, therefore, that two eggs may be in the oviduct at the same time, especially during the height of the laying season.

Structure of an Egg

The accompanying diagram illustrates the structure of a normal egg. Beginning with the outside, an enumeration of the various structures comprises the following:

1. **The shell,** porous to admit air and hard for protection.
2. **Exterior membrane,** a tough membrane calculated to exclude germs and harmful substances.
3. **The inner membrane** immediately surrounding the albumen. This envelops the developing chick and turns with the chick as it pips the shell. The air space at the large end of the egg is between the two membranes.
4. **The albumen,** or white, of the egg, which is formed in concentric layers around the yolk.
5. **Vitelline membrane,** a delicate membrane surrounding the yolk.
6. **Dense layer of albumen.** This is just outside the vitelline membrane.
7. **Yolk,** the yellow layers of albumen just within the vitelline membrane. The yellow color is due to globules of fat. The white yolk is in the center surrounded by the yellow yolk.

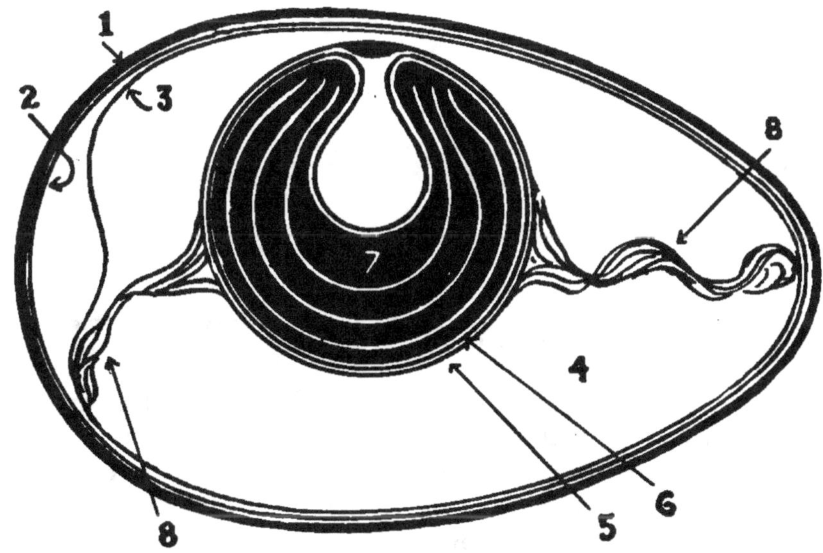

The important parts of an egg are discussed in the text

8. **Chalazæ,** twisted cords of albumen which are attached to opposite poles of the yolk and serve to steady its position in the albumen.

9. **Germinal vesicle,** the living germ cell, or nucleus, which lies upon the surface of the yolk, recognized as a white spot. It may be fertilized or not. As the germinal vesicle lies upon the surface of the yolk it is easily reached by the sperm cells.

Fertility and Fertilization

An egg becomes fertile when the sperm (male) cell fuses with the germ (female) cell. This does not always occur, so that many eggs remain infertile. Where does fertilization take place? Evidently not in the cloaca, nor in the lower portion of the oviduct, for the shell and membranes and albumen would interfere in these regions. Fertilization must take place either at the mouth of the oviduct or in the ovary, possibly in both places. The oviduct is about 18 inches in length; and for the spermatazoa to swim from the cloaca to the infundibulum requires several hours, as this distance must be traversed before they can reach the unfertilized germ cells

It is quite probable that some of the sperm cells find their way into the ovary, as there is evidence that several eggs may be fertilized as the result of a single copulation. In the case of turkeys and geese one copulation seems to be all that is necessary to fertilize all the eggs of a cycle, or clutch. It is possible that the sperm cells remain alive in the oviduct for

Egg types, showing variation in size, also illustrating how the eggs may decrease in size from the first egg of the clutch to the last

a long period and that each germ cell is fertilized as it is discharged into the infundibulum. It is not unlikely that the spermatozoa gain access to the ovary and that all the eggs of the clutch are fertilized very soon after copulation.

How long after a male bird is introduced into a pen before the eggs will become fertile? The sperm cells are provided with slender tail-like filaments, or flagellæ, with which they propel themselves by a whip-like motion, and it is quite probable that some of them reach the mouth of the oviduct within 24 hours. Some eggs, therefore, may be fertile within two or three days. Certainly, in five to 10 days all eggs should be fertile if conditions are normal.

Detecting Infertility and Sex of Eggs

To detect whether an egg is fertile or not before incubation is a problem vital to the poultry industry. After a few days of required temperature fertile eggs show plainly the developing embryo, whereas the infertile egg gives no such evidence. After the infertile egg has been subjected to heat in this manner for several days it loses its freshness and is subject to decay much more than the fresh egg. It is, therefore, unfit for market but may be used for cooking purposes in the home.

As the embryonic cell in a fertile egg differs somewhat in structure from an infertile cell, being enveloped with a distinct ring, it has been thought that fertility and non-fertility could be detected by a powerful lens, but the structural characters are so delicate and the shell and membranes offer such interference to light that this method will probably prove impractical.

It is not impossible that an instrument will yet be devised that will determine not only fertility but the sex of the embryo.

Sex is fixed at the time of fusion of the two sexed cells and it is determined by the chromosomes of the germ cell. Science has accomplished greater things than the problem here presented.

To be able to detect the infertile eggs before subjecting them to the heat of the incubator would be a great boon to the poultry industry. These are the eggs that should be shipped to market, because they have greater keeping qualities than fertilized eggs. It would mean a saving of 25 per cent of all the eggs used for incubation.

If sex could be determined and the poultryman desired to produce more pullets than cockerels, eggs with pullet germs could be selected for the incubator. The theories that sex is developed after incubation, that large and long eggs produce males while small and short eggs produce females, that the first eggs of a clutch usually produce males and those towards the end of the clutch produce females, and that eggs with rough ends produce males while those with smooth ends produce females, may, some of them, have a grain of truth but they cannot be relied upon as infallible.

Abnormal Eggs

We have discussed the development and structure of the normal egg. Under certain unfavorable conditions abnormalities occur. It is well to understand these conditions, as it is possible sometimes to correct them.

A double-yolked egg is formed when two yolks are matured and discharged into the oviduct at the same time. A common coat of albumen, membranes and shell are formed around these making a single egg.

Eggs with blood spots are caused by the rupture of a blood vessel when the follicle cleaves to allow the escape of the yolk. The clot of blood formed escapes into the oviduct and is incorporated in the egg as it is surrounded by the albumen. Such eggs are good for food, as the blood spot does not affect the contents of the egg and may be easily removed.

Bloody eggs arise from some injury, disease or hemorrhage in the oviduct. The blood from the walls of the oviduct becomes distributed through the albumen, and such eggs are not fit for food.

A yolkless egg is formed when some foreign substance gains access to the oviduct through the cloaca. It serves as a

A good home made egg tester

nucleus around which the albumen is deposited and eventually the outer membranes and shell.

A double egg is an egg within an egg and is formed when the original egg by some cause is forced back into the oviduct. A new layer of albumen is formed around it and, as it passes onward, a new shell is formed. Thus we have the strange phenomenon of an egg within an egg, an egg with two layers of albumen and two shells.

Wormy eggs are caused by the passage of worms from the cloaca into the oviduct, where they become enveloped by the albumen before the shell is formed.

Stale fresh-laid eggs, which are foul in odor and offensive in taste, may be traced to individual hens and may be due either to the feed consumed or to a diseased condition of the egg organs.

Gangrene or vent gleet of the cloaca, inflammation of the oviduct, or a disease of the ovary might account for this condition. Probably, however, the retention of the egg in the body of the hen, resulting in incubation and finally decomposition, is the best explanation of this phenomenon, at least this is the most common cause. Sometimes, when deprived of sufficient ash, hens will eat their own droppings, and this might affect the eggs. The remedy is to find the guilty hens and dispose of them.

Incubation

Incubation is the development of the embryo within the egg, and is to be distinguished from fertilization, which is the fusion of the sperm and germ cells. The period of incubation is 21 days, though it may range from 18 to 24 days, depending on the time occupied in supplying the required number of heat units. If the temperature is run at too high a degree, the time of hatching is hastened; if at too low a degree, it is delayed. The normal temperature for incubation is 103° Fahr. Any drop below 90° or rise above 107° is considered dangerous to the hatch. The low temperature is least objectionable.

Eggs will sometimes hatch if allowed to cool for 24 hours in a room of moderate temperature, but such treatment undoubtedly results in weakened vitality. An excess of heat is more disastrous because it coagulates the albumen and hardens the yolk before it is absorbed into the body of the chick, causes the chicks to stick to the shell in hatching, causes many chicks to die in the shell, and produces a batch of weaklings that will never thrive.

Room Temperature for Eggs

As an egg will incubate if kept in a room for any considerable time at a temperature of 70° Fahr. and as the germ will be injured or destroyed if allowed to go below 30°, it is important to know just how to regulate the temperature for eggs intended for incubation. The correct room temperature is 60° Fahr., though a range of five degrees above or below this point will do no harm. A temperature of 50° Fahr. is correct for market eggs.

Size and Weight of Eggs

The standard weight of a dozen market eggs is 24 ounces, and such eggs are of average size. Abnormally large eggs usually have a double yolk. An abnormally small egg may be the first egg of a pullet or the last egg of a cycle, or clutch. When very small, it is devoid of yolk. Eggs vary in weight, even though of the same size, due to the density of the egg contents. The size and weight of an egg are determined to some extent by the breed producing it.

The Minorca lays a white egg, and so large that a dozen will weigh 30 ounces. The Rhode Island Red and Plymouth Rock egg is large and heavy, the average for these breeds being about 26 ounces. In the Vineland, New Jersey, egg-laying contests it was found that the Rhode Island Red eggs were heavier than those of any other dual purpose breed. Leghorns and other Mediterranean breeds lay smaller eggs than do the American breeds, though the White Leghorn may be counted an exception to the rule. This breed has been cultivated for large eggs, so that they compare favorably in size with larger breeds.

The specific gravity of an egg can be tested by placing it in a vessel of water. If the air space is small, the egg contents filling the shell, the egg will lie on its side on the bottom of the vessel. If the contents are well evaporated, leaving a large air cell, the egg may float on the surface of the water. Between these two extremes there are gradations in specific gravity, the determining factor being the size of the air cell. Egg testers are on the market for determining specific gravity. The higher the specific gravity the greater the food value of the egg contents. Only eggs of high specific gravity should be placed in storage.

Color of Eggs

The color of eggs is due to pigments derived from the blood. The pigment is deposited with the carbonate of lime at the time that the shell is formed. The color gradually fades as the hen approaches the end of her laying period. Undoubtedly egg-production has some connection with the fading of the feathers during the laying period. Mottled and variegated eggs occur and arise from the peculiarities of individual hens.

Whether the contents of a white-shelled egg have less nutritive value than the contents of a brown-shelled egg is a subject worthy of investigation. White eggs receive preference in New York city, but in Boston brown eggs are preferred. In the western states an egg is an egg, whatever the color. In marketing, if eggs are graded according to color it adds to the appearance and undoubtedly appeals to critical buyers.

Shape of Eggs

The shape of an egg is determined by the character of the oviduct. The contraction of the oviduct forces the egg onward in its course. This contraction is behind the egg, therefore the small end of the egg is at this point, the large end being toward the outside. If the oviduct is small the tendency will be to make a long egg, if it is large the tendency is toward a round egg. Round eggs are larger in diameter than long eggs, as a rule. Eggs that are abnormal in shape are caused by some abnormal condition in the oviduct.

Round and elongated eggs should never be used for incubation because it is difficult, if not impossible, for chicks to extricate themselves from the shells. Pullets produced from such eggs will lay eggs of similar shape. Such eggs are difficult to pack, the round ones being too large in diameter and the long ones too long for the ordinary container. By careful selection from year to year a uniform type of egg can be established for the flock. Uniformity in product means pleased customers and better prices.

Composition of an Egg

An egg is 66 per cent water and 34 per cent solid or semi-solid matter. The solid matter comprises combinations of the following elements: Carbon, hydrogen, oxygen, nitrogen, sulphur, phosphorus, calcium, sodium, potassium, magnesium, iron, silicon and chlorine. These elements are combined in substances known as protein, ash and fat. The following table shows the percentages of these substances in the whole egg, the white, and the yolk.

Table No. III—Composition of egg.

	Water Percent	Protein Percent	Ash Percent	Fat Percent
Whole egg	65.9	12.83	10.68	10.59
White	87.0	12.00	0.20	0.80
Yolk	50.0	16.00	1.00	33.00

Reference to the above table shows the materials which enter into an egg. What better proof do we need of the importance of providing in the ration all the elements required to manufacture this finished product? If any one of these is lacking the development of the egg is prevented, or at least delayed until the missing material is supplied. As well undertake to build a house without lumber and nails or an engine without iron as to make hens lay without the materials required for the manufacture of an egg. Water is needed, as 66 per cent of the egg is water. Here is where many fail. Eggs cannot be produced in winter without an abundance of water. If the water supply ceases the hens suddenly quit.

Protein is required as it comprises 13 per cent of the whole egg and 16 per cent of the yolk. It must be supplied in a balanced ration. Protein of animal origin seems to be the most helpful in egg production, hence skim milk, dried buttermilk, fish scrap, meat scrap and tankage are recommended to supply this need.

Ash enters into all the structures of an egg, but is especially needed in the shell in the form of carbonate of lime, hence oyster shell is used to supply this need.

Fat constitutes one-third of the weight of the yolk, and as, in the process of egg building, the yolk is the first to be developed we are reminded that a hen must be fat or in good condition before she can engage in productive laying. The fat of her body is appropriated to build the yolk. This suggests the importance of carbohydrates and proteids in due proportion in the laying ration. The reader is referred to Chapter IV, where the subject of feeding rations is fully discussed.

Mr. Osburn feeding his turkeys

Eggs from hidden nests are always open to suspicion. They should always be candled before they are sold.

Chapter VI

Eggs and By-products

ABOUT 35 per cent of all eggs are consumed by the producer. As it is good business to sell the best that customers may be pleased and a reputation for quality established, it is important to cull for home use.

Culling for Home Use

Small eggs should be kept for home consumption. They are as good in quality as the larger eggs, but they reduce the weight of a dozen below standard requirements and are not in favor with purchasers.

Dirty eggs should be kept at home. If freshly gathered their quality is not impaired and, as there can be prompt consumption in the home, they can be washed without harm. whereas washed eggs should never be marketed on account of their poor keeping quality.

Large and elongated eggs should be culled for home use as they are liable to breakage in an ordinary container.

Cracked eggs have such poor keeping quality and are so likely to become leakers and damage many others that they should be kept for home consumption.

Frozen eggs can be used in the household but are not safe to ship.

All eggs of doubtful quality should be kept at home. Such are eggs upon which hens have been allowed to sit for a short period, infertile eggs from the incubator, stale eggs from newly found nests, and bloody eggs or eggs containing blood spots or any foreign matter.

Eggs with blood spots, blood rings, mold and rot can be culled out with an ordinary egg tester. Incubator eggs should never be sent to market because their freshness is destroyed by a few hours of heat, and they are liable to spoil before they reach the consumer.

Keeping, Preserving and Storing Eggs

Infertile eggs keep the best, hence the slogan "Swat the Rooster" (after the breeding season) is a good one. A cool dry cellar is the most desirable place to keep eggs. The temperature should not be allowed to go above 70° nor below 35°, the correct temperature being 50°. The place where they are kept should be clean and sweet. Foul odors injure the quality, and dampness produces mold. Eggs for the incubator should be turned daily and may be kept in cabinets having drawers with movable, slatted bottoms.

Every producer should preserve eggs for future home consumption. November is the lean month, and a few dozen eggs stored for this season of scarcity will be greatly appreciated. Further, such eggs can be sold as storage eggs when the prices are high. The time to do this is in the time of plenty when eggs are cheap.

Three methods of preserving eggs are in use.

1. **Preserving in salt.** This method is very simple and requires only a good tight box with a quantity of salt. The eggs are packed in the salt, a layer at a time, until the box is filled. The eggs for preserving should be clean and fresh. If there is any doubt about their freshness, they should be candled. Place the container in a cool place, and the eggs will keep several months. Infertile eggs for this purpose can be obtained by removing the males from the flock for two or three weeks.

2. **Preserving with lime water.** This method is considered more satisfactory than the salt method. Dissolve about three pounds of unslaked lime in a small quantity of water and then add about seven gallons of water sterilized by boiling. Use a five gallon or a ten gallon crock and, after scalding it, place the eggs in position and pour upon them the clear solution of lime water. Enough water should be added to cover the eggs about two inches.

3. **Preserving with water glass.** This is the most satisfactory method, though more expensive. If a five gallon crock is used it should be cleaned thoroughly and scalded. Then pour into it nine quarts of water that has been sterilized by boiling and to this add one quart of water glass (sodium silicate) and put the eggs in carefully so as not to crack the

shells. The solution should always cover the eggs for one or two inches. When the crock or jar is full it should receive a tight cover and be set in a cool place.

The principle involved in these methods is that of keeping from the egg any bacteria or other harmful substances that might cause decay. Some follow the practice of dipping the eggs first in the water glass solution, and then allowing them to dry. When the water glass sets it effectually closes the pores of the egg. It is then put into the crock and held until ready for consumption. Preserved eggs, intended for boiling, should be punctured with a needle in the large end to allow the escape of the expanding air, else they will crack.

Collecting Eggs

To keep eggs clean change the nesting material occasionally. To prevent cracking or breaking when the eggs are laid use an abundance of nesting material. It is a good plan to cover the bottom of the nest box with sand or clean soil before adding the material. Coarse straw is not very satisfactory. Cut straw or chaff is better. A layer of fine wild hay makes an excellent bed for the nest box. Blind checks and dents are often due to a bare nest box. Hens cultivate the egg-eating habit by reason of broken eggs in faulty nests.

Collect eggs carefully in a vessel kept especially for the purpose. It should be lined with soft material to reduce the danger of breakage. An oval basket answers well for this purpose. In cold weather eggs intended for hatching should be collected twice a day to prevent chilling. The same is true in summer to prevent incubation. If the producer is catering to a private trade and desires fancy eggs, they should be gathered twice daily to prevent soiling. In ordinary weather and conditions, it will be sufficient to collect the eggs once a day.

Egg Losses

The chief losses of eggs on the farm are from the following causes:

1. Neglect in gathering.
2. Incubation by natural heat.
3. Freezing.
4. Keeping in damp or warm room.
5. Loss in the incubator.

6. Breaking, either in the nest or from careless handling.
7. Vermin.
8. Filthy conditions.
9. Careless packing.

All of these losses can be reduced by the producer and, if the carrier, jobber and retailer would cooperate, it would mean the saving of $50,000,000 annually, or a half billion dollars in a decade.

Testing Eggs

Testing or candling eggs is easily accomplished by placing the egg between a strong light and the eye. Testers are made especially for this purpose. The electric tester is probably the most satisfactory.

In candling for market the following defective eggs are detected and rejected:

Checks, eggs with cracked shells.

Blood rings, eggs in which incubation has started but the germ has died.

Broken yolks, eggs in which the vitelline membrane has ruptured and the contents of the yolk are diffused through the albumen.

Spots; these show a dead immovable germ or some foreign substance in the egg or a clot of blood.

Shrunken eggs, which are indicated by the large air cell.

Rots, in which decomposition has set in and the contents appear black.

Eggs intended for hatching should be tested before being set, and all that show the above characteristics should be thrown out.

Testing out Infertile Eggs

It is customary to test eggs under incubation on the seventh to the tenth day and again on the fifteenth to the eighteenth day. The test is made in a warm room. The tray from the incubator is placed on a low table and two baskets are provided, one to receive the infertile eggs and the other for those with dead germs and otherwise defective. Infertile eggs will appear perfectly clear. Dead germs will often be found adhering to the shell membrane. Shrunken eggs will show an excessively large air space. If the air space is filled up the egg is getting too much moisture. Cloudy eggs are defective and indicate a breaking up of the egg contents. Other defects will appear, and with a little practice the operator will soon learn to detect the eggs with strong germs.

The By-Products

It is claimed that the great packing establishments in Chicago make their profits largely on the by-products. Everything is saved and utilized—the hair, blood, viscera, bones and excreta. Some of the manufactured goods are bone meal, granulated bone, meat scrap, meat meal, tankage, blood meal and fertilizer.

The chief by-products of poultry are the feathers and the fertilizer.

Feathers

What is a feather? It is a modified scale, being derived from the skin, or epidermis. The types of feathers are the perfect, the downy and the hairy. The hairy type is illustrated in the hairs that remain on a fowl after plucking and which are usually removed by singeing. Down is the type of feather found in the day-old chick and upon ducks and geese after the outer feathers are removed. The perfect type is developed after hatching and is described below.

The structure of a feather includes the following parts:

Quill, the naked hollow barrel which is inserted into the skin.

Shaft, which is a continuation of the quill. It is rectangular in cross-section.

Vane, the expanded portions of the feather attached to the sides of the shaft.

The vane consists of the **barbs,** which are branches from the shaft; the **barbules,** branches from the barbs; and **barbicels,** branches from the barbules. The barbules and barbicels have **hooklets** by means of which these structures are interlocked. Thus, by these overlapping and interlocking parts, the vane, or web, is formed. This makes the feather very resistant to air in flight.

Composition of a feather. The feather is composed of a large percentage of silica. The organic constituent consists approximately as follows: Carbon, 50 per cent; hydrogen, 6 per cent; nitrogen, 17 per cent; oxygen, 23 per cent; sulphur, 4 per cent.

The abundance of nitrogen in the composition suggests the importance of a nitrogenous diet during the molt.

Molting of Feathers

This is the shedding of the feathers and the growth of a new supply. Some birds molt twice a year, in the fall when they put on their winter clothes preparatory to cold weather and in the spring when they put on their breeding dress.

As a rule, however, there is but one molt, in the fall of the year, the change in plumage which occurs in the spring being but an increase in coloring matter. During the winter the plumage is often of a protective nature, resembling the somber colors of that season. Chickens and other domestic fowls undergo a complete molt in the summer or fall.

The growth of new feathers makes a severe drain upon the vitality of a fowl. It must have a very important bearing upon egg-production, for when a hen molts the surplus food is required to make feathers, and egg-production naturally ceases. Some hens molt so gradually and the draught upon the vital forces is so moderate at any one time that they keep on laying through a portion of the molt. Others seem to shed their feathers in a day. Then follows a long period in which new feathers must be grown, and that means a period of rest from egg-production.

It is claimed that the early molter is the poor layer. This is not necessarily so. To illustrate, a good laying pullet may continue laying until early summer. She becomes broody and, if allowed to sit too long, she loses flesh, in which condition she is not able to resume egg-production. If such a bird is broken of her broodiness and suddenly put on a heavy nitrogenous diet she may be forced into an early molt. To cull her from the flock for this reason would be unjust.

It has been proved that the period of molting can be controlled by the system of feeding. Hens that are put on a starvation diet for two weeks and then fed heavily will often molt out of season, even in midsummer. When the egg-ration is withdrawn they cease to lay, when they cease laying they begin to molt. It is also probable that the exact time of molting is affected by the time of hatching. Pullets hatched very early usually molt in the fall. I have known March hatched cockerels to molt in October. Pullets hatched late may miss the fall molt but will molt the following summer and yet be very good layers. To prevent the summer molt it is necessary to provide an egg-producing ration, so that the hens may be kept laying as long as possible.

Late molters are good layers. This is a very safe rule. They lay more eggs than the early layers because the laying season continues through a longer period. If a hen molts late she will have a shorter rest period. Feathers shed during the molting season are lost for commercial purposes. The time required to produce a complete new dress is about three months.

Uses of Feathers

Here are some of the commercial uses of feathers:

They are used for the manufacture of pillows, beds, cushions, ornaments for apparel, toys, dusters, fans, muffs, feather bone, feather board and brooders. In Three Oaks, Michigan, is a large factory which utilizes the tail feathers of turkeys for the manufacture of featherbone and feather board. These articles are used in making corsets and other apparel for women. Feathers are always in demand and meet ready sale if properly prepared.

Preparing Feathers for Market

Dry-picked feathers are preferred to those plucked after scalding. Scalded feathers must be thoroughly dried before shipping. White feathers command a better price than colored. The most valuable are the down and fine feathers of geese. The fine feathers of chickens also command a good price, especially if dry-picked. Feathers should be graded for market and the different grades packed separately.

In picking turkeys, save all the feathers that grow on the tail, also those on the two joints of the wing next the body. The pointed, one-sided feathers, or primaries, that grow on the end of the wing sell at a low price and should be kept separate from the others.

In packing, lay quill feathers straight in as light boxes as possible. If stuffed into bags they become broken. Body feathers should be shipped in burlap sacks. Chicken feathers should be forked over to allow the animal heat to escape and to assist in drying. Dampness mats them together and in that event they are liable to arrive at market in a heated or moldy condition. If quill feathers are mixed in with the body feathers it will mean a cut in the price.

Fertilizer

Poultry manure is rich in nitrogen and hence is a very valuable fertilizer. It is usually used on the garden, where a very rich soil is necessary. Poultry manure loses strength rapidly when piled outdoors. It should be spread on the soil as soon as possible. Do not mix poultry manure with wood ashes or lime, as this causes a rapid loss of nitrogen.

A tight dropping board under the roost saves the droppings and keeps the floor clean. It is a good thing to sprinkle the dropping board with some good absorbent, such as peat, sawdust or sifted coal ashes. The dropping board should be cleaned off once or twice a week.

Many millions of dollars are lost annually by failure to utilize the by-products of poultry. The wise poultryman will gather up the crumbs that nothing may be lost.

Chapter VII

The Puzzle of the Breeds

THERE are 149 varieties of land and water fowl recognized by the American Poultry Association. Of these. 121 are chickens, 15 are ducks, seven are geese and six are turkeys. These varieties represent 60 breeds and 15 general classes. Who can measure the thought and patience and skill required to produce these results? Scientific work of a high order, carried on through many years, has been necessary, and much credit must be given constructive breeders for the service they have rendered.

What is a Breed?

The term **class** is applied to a group of breeds having a common origin and a close resemblance in type. To illustrate, the Asiatic class is represented by Cochins, Brahmas and Langshans, breeds that have originated in Asia, and there is a resemblance in structural characters.

A breed is a group of individuals having a common type. Breed is determined more by shape than by color. There may be several varieties of the same breed. For example, the Plymouth Rock breed which belongs to the American class, embraces six varieties, but all of these have a common resemblance in shape.

A variety is a group of individuals which have common color patterns as well as the shape characteristic of the breed. The White Plymouth Rock is a variety of the Plymouth Rock breed.

A strain is a group under a variety. It is produced by inbreeding in the hands of the expert breeder. The Hawkins strain of Barred Rocks, the Tompkins strain of Rhode Island Reds and the Barron strain of White Leghorns are good illustrations. A strain comprises individuals which excel in shape. color or egg-production, and this is brought about by the skill of the breeder.

How to Select a Breed

A purebred fowl is one that breeds true to breed characteristics. There may be an occasional exception, but this is due to reversion, the fowl breeding back to some ancestor. If mated differently this same fowl might breed true. This tendency to revert is overcome by inbreeding.

Is the Purebred Worth While?

Every breed has points of excellence. If it were not so, it would have perished in the making. Is the farmer better off with a purebred flock? In this connection let us note:—

1. That it costs no more to feed a purebred than a mongrel.
2. That the purebred is more attractive and will command greater interest and receive better care than the mongrel, and will therefore prove more profitable. In the development of a purebred, vigor and high production are emphasized as well as fine feathers, so it will be more productive than the mongrel, which, on account of promiscuous breeding, will be found wanting.
3. The purebred will produce a uniform product in carcass and egg, and this means an appeal to the purchaser, insuring a ready market and top prices.
4. There are sources of income from a purebred flock not found with the mongrel, such as eggs for hatching, day-old chicks, and breeding stock.
5. When farms are located near each other and a farmer has his separate breed, he can indentify his own stock in case of accidental mixing.
6. Purebred fowls create a special interest among young people, appealing to the nobler sentiments. This benefit alone is worth all the extra expense in establishing the flock.

It is a matter of great interest and encouragement that so many purebred flocks are being established on the farms. It is a safe estimate that more than 50 per cent of the flocks in the states of leading production are comprised of standard bred fowls. The tourist along country highways cannot fail to be impressed with the growing interest in purebred poultry.

What is the Best Breed?

It is commonly stated that one breed is as good as another. This statement should not go unchallenged. There is a best breed for the individual, the environment, the end to be attained, and the location of the farm or plant.

The best breed for the individual must be determined by his own likes and dislikes. If he has a nervous temperament he will undoubtedly choose an Orpington in preference to a nervous Leghorn. If bright colors appeal he will probably overlook the Black Langshan. It is the appeal to the individual that usually determines the breed selected. It would be unwise to select a breed toward which there is any aversion. The question of environment may determine the best breed to be chosen. For a severely cold climate a Light Brahma would have preferences over a Leghorn. A hot climate would not be suited to the Buff Cochin. If the soil is damp and poorly drained and surroundings unfavorable, a breed of great physical stamina would be the best.

Single comb Rhode Island Red cockerel

The end sought should have some influence in determining what is the best breed. If market fowls are desired it would be unwise to select the Leghorn, Hamburg, or Andalusian, but a meat breed should be selected. If eggs are the aim, then an egg breed is best. The location with reference to market should be a determining factor. A breed producing white eggs would be indicated for locations near New York City and San Francisco, but near Boston and other localities, the breed of the brown egg would be the best. There is a best breed. How to determine it is not always an easy task.

How to Select a Breed

Having considered well the soil, the climate and the location, the first thing to determine is the character of breed that will meet the conditions and fulfill the aim of the producer. Breeds are classified as meat breeds, dual-purpose breeds, egg breeds and fancy breeds. A meat breed is one whose tendency is to lay on flesh. Such breeds are large, docile, good feeders, but, as a rule, poor layers. The egg breeds have been trained for egg-production by years of careful selection. They are usually small in size, not heavy eaters, do not fatten readily, and make poor market fowls. They have the faculty of turning the food into eggs and this is done at less expense than by the heavier breeds. They are as a rule nervous and excitable, but can be trained to become very friendly. The dual-purpose breeds partake of the qualities of the two classes just defined.

The leading dual-purpose breeds are of American origin. They range between the meat breeds and the egg breeds in size. They make choice table fowls and at the same time are good egg-producers. The fancy breeds are those cultivated because of certain marked peculiarities of shape and color. They appeal to the fancy of the breeder and are bred as a matter of interest and pleasure, rather than as a source of profit.

Which one of these classes shall the poultry-keeper select? The average farmer will select the dual-purpose breed. The commercial poultryman, whose plant is near the great city, will select an egg breed if he wishes to cater to the market for choice eggs. If his aim is to produce the finished carcass, he will select a heavier breed. The amateur, whose aim is to satisfy his interest in the fancy, will select a breed of that character.

But there are several breeds of each class and it is sometimes very difficult to make a definite choice. If the dual-purpose class has been chosen, the individual must decide whether to adopt the Plymouth Rock, Wyandotte, Rhode Island Red, Orpington, or another of the several general-purpose breeds. Having decided on the breed, there remains the still difficult task of selecting the variety. This may be determined by his own fancy or by his knowledge of some good variety, or by the variety of some good breeder in whom he has confidence and whose advice he is willing to consider.

After the variety is chosen then he must decide upon the strain, for he knows well that if he would make rapid progress he should secure his foundation stock from an established strain. If he does not do this he will have to build a strain of his own and this will require years of patient effort.

At least 75 per cent of all purebred flocks are on the farms, and farmers not only furnish the demand for purebred stock produced by the exclusive fancier and are the means of perpetuating existing breeds, but many of them are constructive breeders and are to be credited with producing new breeds or making improvements on those already established.

Standard of Perfection

It would be unwise to undertake in the limited compass of this book even a brief description of all the standard varieties of poultry. For full and accurate descriptions of all recognized breeds the reader is referred to the "Standard of Perfection" published by the American Poultry Association. This excellent book can be secured from Prairie Farmer's Book Bureau. Every farmer or fancier who is building a purebred flock will receive great help by securing a copy of this book.

Key to Poultry Breeds

The following key to the recognized poultry breeds is acknowledged to be brief and incomplete. The descriptions are only suggestive, but may help the beginner in his selection of a breed and in identifying at least the more common breeds.

Class I. American Breeds: Dual-purpose breeds of American origin, medium size, clean legs.

Breed No. 1.—Plymouth Rock: With long, broad and deep body with full breast. Cock, 9½ lbs.; hen, 7½ lbs.

Varieties:

(1) All feathers barred white and dark. .... Barred Plymouth Rock
(2) Plumage pure white in all sections...................... White Plymouth Rock
(3) Plumage rich buff in all sections......................... Buff Plymouth Rock
(4) Black and white with pencilings................ Silver-Penciled Plymouth Rock
(5) Red, brown and black, with pencilings............... Partridge Plymouth Rock
(6) White, with black in hackle, wings and tail.......... Columbian Plymouth Rock

Breed No. 2.—Wyandotte: Body short, broad, deep and round. Rose comb. Cock, 8½ lbs.; hen, 6½ lbs.

Varieties:

(1) Pure white in all sections.............................. White Wyandotte
(2) Pure buff in all sections................................. Buff Wyandotte
(3) Greenish black in all sections........................... Black Wyandotte
(4) Black with silver lacing................................. Silver Wyandotte
(5) Black with golden lacing................................ Golden Wyandotte

(6) Black and white with pencilings....................Silver-Penciled Wyandotte
(7) Red, brown and black with pencilings..................Partridge Wyandotte
(8) White, with black in hackle, wings and tail............Columbian Wyandotte

Breed No. 3.—Javas: Single comb, long back. Cock, 9½ lbs.; hen, 7½ lbs.
Varieties:
(1) Black with greenish sheen..Black Java
(2) Mottled black and white....................................Mottled Java.

Breed No. 4.—Dominiques: Rose comb; back with concave sweep, medium length and broad. Cock, 7 lbs.; hen, 5 lbs.
Varieties:
(1) Irregular bars of slate and white......................American Dominique

Breed No. 5.—Rhode Island Red: Back broad and long, breast full, body oblong, closely feathered. Cock, 8½ lbs.; hen, 6½ lbs.
Varieties:
(1) Rich red, black in tail and wings, rose comb......Rose Comb Rhode Island Red
(2) Single comb...............................Single Comb Rhode Island Red

Breed No. 6.—Buckeye: Broad and long back, rounded breast. Cock, 9 lbs.; hen, 6½ lbs.
Variety:
(1) Color mahogany bay, pea comb..........................American Buckeye

Class II. Asiatic breeds: Legs and toes feathered, large size, meat breeds.

Breed No. 1.—Cochins: Round and plump body, single comb. Cock, 11 lbs.; hen, 8½ lbs.
Varieties:
(1) Prevailing color buff...Buff Cochin
(2) Plumage white ....White Cochin
(3)Plumage black...Black Cochin
(4) Prevailing colors black and red............................Partridge Cochin

Breed No. 2.—Brahmas: Body long, compact, closely feathered. Pea comb. Cock, 12 lbs.; hen, 9½ lbs.
Varieties:
(1) White, except black in hackle, wings and tail................Light Brahma.
(2) Plumage black and white....................................Dark Brahma

Breed No. 3.—Langshans: Short body with concave sweep to back; single comb. Cock, 9½ lbs.; hen, 7½ lbs.
Varieties:
(1) Plumage black..Black Langshan
(2) Plumage white..White Langshan

Class III. Mediterranean breeds: Body slender and small, graceful carriage. Legs and toes clean. Egg breeds.

Breed No. 1.—Leghorns: Moderately long body, yellow legs, white ear lobes; sprightly carriage.
Varieties:
(1) Plumage white, rose or single comb.........................White Leghorn
(2) Plumage brown and black, rose or single comb..............Brown Leghorn
(3) Plumage buff, rose or single comb............................Buff Leghorn
(4) Plumage black, single comb..................................Black Leghorn
(5) Silvery white, with black in hackle, single comb..............Silver Leghorn
(6) Orange, red, salmon, white..............................Red Pyle Leghorn

Breed No. 2.—Minorcas: Large size, long body. Cock, 9 lbs.; hen, 7½ lbs.
Varieties.
(1) Plumage black, rose or single comb.........................Black Minorca
(2) Plumage white, rose or single comb.........................White Minorca
(3) Plumage buff, single comb....................................Buff Minorca

Breed No. 3.—Black Spanish: Plumage black; ear lobes and face white. Cock, 8 lbs.; hen, 6½ lbs..White-Faced Black Spanish

Breed No. 4.—Andalusians: Leghorn type, medium size, graceful carriage. Cock, 6 lbs.; hen, 5 lbs. Plumage blue.....................................Blue Andalusian

Breed No. 5.—Anconas: Leghorn type. Cock, 5½ lbs.; hen, 4½ lbs.
(1) Plumage mottled black and white............................Mottled Ancona

Class IV. English Breeds: Dual purpose breeds, excelling in meat qualities. Medium to large size.

Breed No. 1.—Dorkings: Body broad, oblong, low-set; five toes; single comb. Cock, 7½ lbs.; hen, 6 lbs.

Varieties:

(1) Plumage white ....White Dorkings
(2) Colors white, black and salmon........................Silver Gray Dorking
(3) Colors black, straw, salmon, gray; 4-toes.....................Colored Dorking

Breed No. 2.—Redcap: Body long, cylindrical. Rose comb.

(1) Predominating colors, red and bay........................English Red Cap

Breed No. 3.—Orpingtons: Large size. Body long, broad and deep. Legs and skin white. Cock, 10 lbs.; hen, 8 lbs.

Varieties:

(1) Plumage white in all sections............................White Orpington
(2) Plumage buff in all sections...............................Buff Orpington
(3) Plumage black in all sections............................Black Orpington
(4) Plumage blue in all sections...............................Blue Orpington

Breed No. 4.—Cornish: Broad breast and back, upright carriage, pea comb, stout legs. Cock, 10 lbs.; hen, 7½ lbs.

Varieties:

(1) Plumage greenish black.....................................Dark Cornish
(2) Plumage white...White Cornish
(3) Plumage red, with white lacing.....................White-laced Red Cornish

Breed No. 5.—Sussex: Body oblong; legs, skin and flesh white. Cock, 9 lbs.; hen, 7 lbs.

Varieties:

(1) Reddish brown, white speckled............................Speckled Sussex
(2) Mahogany red...Red Sussex

Class V. Polish Breeds: Moderate size, prominent crest.

Breed No. 1.—Polish: Body medium length, tapering toward rear. Pea comb. Moderate size.

Varieties:

(1) White crest ....White-crested Black Polish
(2) Golden crest and beard............................Bearded Golden Polish
(3) Plumage white, with black lacing......................Bearded Silver Polish
(4) Plumage white...Bearded White Polish
(5) Plumage buff, laced with pale buff.......................Buff-laced Polish
(6) Without beard................................Golden Non-bearded Polish
(7) Without beard................................Silver Non-bearded Polish
(8) Without beard ....White Non-bearded Polish

Class VI. Hamburgs: Small size, brilliant colors, white ear lobes, blue shanks.

Breed No. 1.—Hamburgs: Body round, prominent breast, rose comb.

Varieties:

(1) Golden bay, with black spangles..................Golden Spangled Hamburg
(2) Silvery white, with black spangles................Silver Spangled Hamburg
(3) Reddish bay, penciled black.......................Golden Penciled Hamburg
(4) Silvery white, penciled black.....................Silver Penciled Hamburg
(5) Pure White...White Hamburg
(6) Greenish Black ....Black Hamburg

Class VII. French Breeds: Dual purpose breeds, excelling in table qualities; V-shaped comb.

Breed No. 1.—Houdans: Crest and beard, toes five. Cock, 7½ lbs.; hen, 6½ lbs.

Varieties:

(1) Plumage black and white................................Mottled Houdans
(2) Plumage white..White Houdans

Breed No. 2.—Crevecoeurs: Back broad, breast full, body compact, legs short. Cock, 8 lbs.; hen, 7 lbs.

(1) Plumage black, crest present...............................Crevecoeur

Breed No. 3.—La Fleche: Without crest. Large and powerful body. Cock, 8½ lbs.; hen, 7½ lbs.

(1) Plumage, glossy black..La Fleche
Breed No. 4.—Faverolles: Single comb and with beard and muffs. Cock, 8 lbs.; hen, 6½ lbs.
(1) Colors, salmon, brown, and black........................Salmon Faverolles

Class VIII. Continental: Egg breeds, active, productive.

Breed No. 1.—Campines: Long body, full breast, single comb. Cock, 6 lbs.; hen, 4 lbs.
Varieties:
(1) Greenish-black barred with white............................Silver Campines
(2) Greenish-black barred with golden bay......................Golden Campines

Class IX. Games and Game Bantams: Comprise two breeds and sixteen varieties.

Class X. Orientals, including **Black Sumatras** and **Malays,** three breeds and three varieties.

Class XI. Ornamental Bantams, eight breeds and nineteen varieties.

Class XII. Miscellaneous Breeds, including **Silkies, Sultans** and **Frizzles.**

Class XIII. Ducks.

Breeds	Varieties
Pekin	White
Aylesbury	White
Rouen	Colored
Cayuga	Black
Call	Gray
Call	White
East India	Black
Muscovy	Colored
Muscovy	White
Swedish	Blue
Buff	Buff
Crested	White
Runner	Fawn and White
Runner	White
Runner	Penciled

Class XIV. Geese.

Breeds	Varieties
Toulouse	Gray
Emden	White
African	African
Chinese	Brown
Chinese	White
Wild	Gray
Egyptian	Colored

Class XV. Turkeys.

Breed	Varieties
Turkeys	Bronze Narragansett White Holland Black Slate Bourbon Red

Every breed in this remarkable list has a history. It may be that all chickens hark back to a common ancestry, and the same may be true of ducks, geese and turkeys. Under the laws of selection (natural or artificial), variation, and prepotency original characters have been changed and by inbreeding new types have been established. Much of the past history of every breed is covered from human knowledge, so that our knowledge of the origin of any breed or variety is very limited. The near-origin of some of the breeds is quite well known. It may be helpful to cite a few of the more popular breeds, giving their supposed origin and some of their valuable characteristics.

Origin and Value of Popular Breeds

To know the origin of a breed puts the poultry-keeper in possession of the good and bad characters of the ancestral breeds and enables him to conduct his breeding operations in such a way as to eliminate the undesirable qualities. To know the points of value in the breed of his choice enables him in the care and feeding of his flock to emphasize and improve these qualities. There are other breeds than those named that are just as good from a utility or aesthetic standpoint as those given, but these are the breeds that seem to the writer to be most common on the farms. They probably represent 90 per cent of all farm poultry.

Barred Plymouth Rock Pullet, Oblong type

Barred Plymouth Rocks

This is pre-eminently the farmers' breed, though its popularity has waned in some degree since the appearance of other breeds that vie with it in excellence. As to its origin, it is claimed to have been produced from the American Dominique as a foundation, and upon this foundation have been placed several different courses of breeding. In this view the Barred Rock can have more than one origin, but the consensus of opinion is that Asiatic breeds crossed upon the Dominique are the source of the modern Barred Rock. Dark and light specimens appear in the breeding, and this necessitates a

Pullet Bred Barred Plymouth Rock Cockerel

double system of mating as necessary in order to produce specimens of show-bird excellence. In recent years the two sub-varieties are judged separately, and there are some breeders who breed one type to the exclusion of the other. The tendency of the female is to become darker in color and of the male to become lighter. This prevents uniformity and the fault is overcome by double mating.

The Barred Rock is a splendid dual-purpose breed. The chicks grow rapidly and are thrifty. The flesh is highly prized. They are good layers in all seasons if fed for egg-production. The hens are good sitters and mothers. Eggs are large and heavy and, as the flesh surpasses in tenderness and quality, they are in great demand in the markets. The pictures on this and the preceding page show a typical cockerel and pullet of this popular breed.

White Plymouth Rock

This variety originated as a sport from the Barred Rock, and its color was fixed by inbreeding. White Rocks have made a strong appeal to the farmer, and it is quite a common thing to see large flocks. They make an attractive picture on the lawn or meadow. The development of this variety has been in good hands, so that strains, such as the Fishel and Halbach, have been developed which excel as market specimens and surpass in egg-production. At the national egg-laying contest at Mountain Grove, Mo., Lady Show-you, a White Rock hen, attracted world-wide interest by defeating all entrants by laying 281 eggs in twelve months. In keeping the breed pure the disqualifications to be avoided are side-sprigs on comb, feathers on legs, white in ear-lobes, or black in plumage.

Columbian Wyandotte Pullet showing the round type of body characteristic of the Wyandotte

White Wyandotte

The Silver Wyandotte is of American origin, several breeders having contributed toward its production. French, Hamburg and Asiatic breeds were used. The White Wyandotte is a sport from the Silver Wyandotte. The plumage is pure white, ear lobes red, and shanks yellow. The comb is rose. A single comb or a foreign color in any section is to be avoided. This breed has a host of enthusiastic admirers. It is famous for egg-production and the quality of the meat produced probably excels that of the Plymouth Rock. It is represented in some of the large commercial egg-farms and is a common breed on the farm.

Rhode Island Red

The Rhode Island Red embraces two varieties, the Rose Comb and the Single Comb. These are alike in all respects except the comb. The Rhode Island Red is of American origin and takes its name from the state where it is believed to have originated. The breeds entering into its composition are the Cochin China, the Black-breasted Red Malay, and the Brown Leghorn.

On account of its composite character, it is a breed difficult to breed true to shape and color. By the law of reversion there is a constant tendency to throw specimens that are off in type or color. It will require time to breed out this tendency. Marked improvement in the breed has been made in recent years. The Rhode Island Red is a hardy fowl, excelling in winter egg-production and is also an excellent table fowl. For these reasons the breed is in great demand.

The chief objection to it is the inclination to broodiness during the summer months. As sitters and mothers the hens excel. Chicks make rapid growth and choice broilers are made at an early age. Pullets mature in seven months. Eggs are large and smooth in outline, the number of abnormal eggs being remarkably few. This breed is giving the Barred Rock a close race for supremacy on the farm.

Buff Orpington

All varieties of Orpingtons were originated by William Cook of Orpington, England. They are good general purpose fowls, excelling in the quantity and quality of meat which they furnish. In egg-production they do not equal the American breeds, but they are considered very good winter layers. The hens are good sitters and mothers; indeed they are so persistent in sitting that this habit is counted against them. The legs, skin, and meat are white. The white flesh is objected to in some localities, but this unwise prejudice will disappear in time because of their real merit as market fowls. Orpingtons have a strong tendency to have feathers on toes and shanks because of their origin. This disqualification must be guarded against.

The White Orpington is almost as popular as the Buff. It has the same excellent qualities, but the color is white.

Single Comb Buff Orpington hen, wedge type to round

Light Brahma

Of the meat breeds probably the Light Brahma is bred more extensively than any other. It is of Asiatic origin. It has been used to good advantage in forming some of the American breeds. It is a handsome breed, excelling in meat production, and is a fair layer. It is noted for vigor and prepotency, and is docile, and for that reason may be kept easily in confinement. Pullets develop to laying maturity in about ten months.

Black Langshan

The Black Langshan is a handsome breed and not uncommon on the farm. It is a good layer, the skin and flesh are white. As a table fowl it stands in good repute. It is considered a hardy breed. Like the Brahma it is of Asiatic origin.

The White Leghorn

This breed is of Italian origin. Since its introduction into the United States it has been so wonderfully improved that the original stock suffers by comparison. English breeders come to this country for their supply of birds. The White

White Leghorn Hen, Oblong type

Leghorn, as found in America, is supreme in egg-production. Its popularity is evident in every state, but on the Atlantic and Pacific coasts where the great commercial egg factories are found, this breed is most abundant.

Mottled Ancona

The Ancona is a breed of Italian origin and is being recognized as one of the greatest of egg-machines. It is more or less abundant on the farms, and is certain to increase in popularity in all sections where egg-production is of prime importance. The egg is white, of good size, and is produced economically. The general color of the breed is black with about every fifth feather tipped with white.

Campines

Silver and Golden Campines have been bred in northern Europe for many years. They have the egg-type of the Mediterranean breeds and are believed to have come originally from the shores of the Mediterranean. The general color is white with the feathers barred with distinct bands of white or golden bay, according to the variety.

They are hardy, vigorous, great hustlers and wonderful egg-machines. They are gaining in popularity in this country on account of their remarkable foraging habits, subsisting where other fowls would perish.

DUCKS

Duck culture is quite general on the farms of the United States, though on a limited scale. There is the meat type and the egg type. In some sections duck culture is carried on very extensively, as in the Long Island district.

Pekin Duck

This breed was introduced from China and is now well distributed throughout the United States. It is the most popular meat breed and is used almost exclusively on the great commercial duck farms of the East. Squabs are sold at eight to ten weeks of age, when they weigh 4½ lbs. to 6 lbs. The Pekin has a long, broad and deep body and the plumage is creamy white. A black bean in drake or foreign color in plumage disqualifies.

The drake weighs 9 lbs. and the duck, 8 lbs. This duck lays about 100 eggs in a season, and the fertility is usually good.

The Aylesbury Duck

This duck is of English origin and is probably next to the Pekin in utility and popularity. It differs from the Pekin in having a horizontal carriage instead of upright, the beak is pale, flesh-colored instead of orange-yellow, the shanks and toes are light orange instead of reddish orange, and the plumage is pure white instead of creamy white. The weights are same as those of the Pekin. It is a good layer, a fine market duck, and well adapted to commercial farming.

GEESE

Six breeds of geese are recognized, Toulouse, Emden, African, Chinese, Canadian or Wild, and Egyptian.

The Gray Toulouse, named from the city of Toulouse in South France, is extensively bred on account of its large size and good market qualities. The goose is a good layer, producing 20 to 35 eggs in a season. The color is gray, shading into white on abdomen. The gander weighs 26 lbs. and the goose, 20 lbs. The sexes resemble each other, but can usually be distinguished.

Emden Goose

The Emden goose is a native of Germany, taking its name from the city of Emden, which was the central market for the geese of a large district. The gander weighs 20 lbs. and the goose 18 lbs. This is a hardy, popular and profitable breed.

TURKEYS

There are six varieties of turkeys, all originating from the Wild Turkey of America. They still retain their wild instincts, but some varieties are more domestic than others.

The **Bronze** turkey is the most popular of all breeds on the farm. The general color is bronze, but in some sections are black and white bars. The tom weighs 30 lbs. and the hen 18 lbs. It is extensively bred, almost to the exclusion of the other breeds.

The Bourbon Red is probably next in popularity to the Bronze. It is more domestic than the Bronze and the flesh is considered of finer quality. This variety has replaced the

Buff turkey in the standard because it is bred more abundantly and is bred true to color with greater ease. The Buff turkey was difficult to breed true to standard requirements, but, on account of its domestic habits, rapid maturity, and fine table

White Holland Turkey Fowl

qualities, is still retained by a number of breeders. The color of the Bourbon is a rich dark bay, the wings and tails being white. The adult tom weighs 30 lbs. and the hen 18 lbs.

The White Holland is a sport from the other breeds. It is now bred to large size by infusion of blood from Bronze sports. It is especially valued as a market fowl, and the feathers are much prized.

Chapter VIII

The Breeding Problem

BREEDING is a process by which a race is established, improved and perpetuated. This process may be natural or artificial. Natural breeding requires long periods of time to fix the characters of a race so that it can become entitled to specific rank. Such a race, when once established, becomes fixed and permanent and reproduces its kind. The Mourning Dove may illustrate. Its characters have become fixed by natural inbreeding. It is a species. Its sexes do not pair with other species.

In artificial breeding, however, the human element enters as a controlling influence. The selection is artificial. Man selects the foundation stock, mates the sexes, and determines the environment. He follows a system of inbreeding that fixes the type. The result may be the Silver Campine or the Golden Campine. These varieties of chickens have a right to specific rank just as the Golden-winged Woodpecker and the Red-shafted Flicker which closely resemble in type but differ in color.

The difference between the Flicker and the Campine is that the former is more permanent. The Campine persists as long as its destiny is presided over by the mind of man. As soon as his guiding hand is removed the species relapses. Then follow cross-breeding, interbreeding, and a rapid retreat to the jungle type. This might not occur if the group could be kept isolated from other varieties of chickens, and, in that event, it is probable that in time the species would become so fixed that there would be no attempt toward interbreeding. But, if the guiding hand of man should be withdrawn and the breeds of chickens were all thrown together, the 300-egg hen would soon disappear. All the beauty and utility acquired by the thought and patience of long years of endeavor would be quickly destroyed. Man's controlling influence must continue, to retain what has been gained. To establish and maintain a race with desirable characters is the aim of breeding.

The fundamental upon which all breeding rests is reproduction.

Reproduction

Reproduction is the process by which the individual perpetuates its kind. This is accomplished by producing new individuals. There are two kinds of reproduction, known as **asexual** and **sexual.**

Asexual reproduction may occur with the higher forms of plant life, as in the case of grafting and slipping, where there is no evidence of sex interference. As a rule, asexual reproduction is found among the lower forms of life.

Sexual reproduction occurs when one or two sexual cells unite to reproduce a plant or animal. Among plants the sexual elements are the spermatazoids and the oöspheres. The sexual elements of animals are the spermatazoa and the ova. The varieties of sexual reproduction are conjugation, parthenogenesis, and fertilization.

(a) Conjugation consists in the union of two similar cells. In this method the cells are structurally the same, but as the process is analagous to the sexual method—cytoplasm fusing with cytoplasm, and nucleus with nucleus—it is considered by the best authorities as sexual in character. Examples are found in diatoms and animalcules.

(b) Parthenogenesis, or unisexual reproduction, is accomplished by the female. A single sexual cell may be concerned or two cells of the same sex. Illustrations are found in the honey bee, where drones are produced by unfertilized eggs from the queen; also in the plant louse, whose female may go on producing broods of individuals for ten or more generations without the intervention of the male. Then a male appears, and reproduction by male and female follows.

(c) Bisexual reproduction, or fertilization, requires both male and female and is illustrated in all species of birds. Birds reproduce by means of an egg, or ovum, which is produced by the female. The egg is really an enlarged cell, being much enlarged to provide room for the storage of nourishment. The real center of life in the egg is found on the surface of the yolk and is known as the **germinal vesicle.**

Fertilization occurs when the spermatozoön, or sperm cell, fuses with the germinal vesicle. This cannot take place until after the spermatazoa gain access to the cloaca. This takes place in copulation or it may be brought about artificially. After entering the cloaca, the spermatazoa swim to the infundibulum in the fluid secretions of the oviduct, and here is where fertilization takes place. The sperm cells collect at this point in numbers, so that when the yolk is discharged into the mouth of the ovary they are ready to penetrate the germinal vesicle.

They must have wonderful vitality and be capable of living many days, for hens have been known to produce fertile eggs for upwards of two weeks after the male has been removed from the pen. It is at the time of fertilization that sex is determined, not after the egg is laid, as some suppose, and the determining factor is the chromosomes. In the initial cell formed by the union of the two parent cells reside all the possibilities of the individual. Color, shape, temperament, prepotency, and productive power are all wrapped up in this miscroscopic point.

Laws of Reproduction

Some of the laws of reproduction are stated below:

1. **Like produces like.** A Wyandotte will produce a Wyandotte. A single comb will produce a single comb. If this law did not prevail all efforts to produce and improve new breeds would be useless.

2. When two individuals of divergent characters are interbred, some of the progeny will resemble the sire, others the dam; some will show characters that hark back to the grandparents, some will exhibit variations and a blending of the characters of sire and dam. These facts enable us to understand why two unrelated birds of the same variety, though closely resembling in type and color, will produce specimens quite different in these respects from the parents. If we knew fully the history and peculiarities of the grandparents we could prophesy more accurately regarding the characters which would appear in the offspring. This also explains why a rose comb bred with a single comb will result in combs partaking of the peculiarities of both; why a white bird bred with a black will result in a mingling of the two.

3. **Atavism.** While we are confident that like produces like, we must not forget that there are certain influences which seem to antagonize this law. Environment is one of these influences. Modifications in the color pattern of some birds

can be traced to climatic influence. So, also, injury or dietary deficiency might account for foreign color in the plumage, where none is found in the ancestry. Another factor is discordant elements that may have been introduced into the blood somewhere along the line of ancestry. These crop out and the law stated in paragraph one seems to be contradicted. It is not, however, for the blood contains the very elements which appear in the breeding. This tendency to revert to the characters of the original ancestry is known as atavism, or reversion, or "throwing back." I have in my flock of Reds a Brown Leghorn pullet which was hatched from an egg of one of my purebred hens. There has been no Brown Leghorn on the farm for 12 years. This pullet seemed to drop out of a clear sky. The cause is apparent. The Brown Leghorn was one of the foundation breeds of the Rhode Island Red. By the law of reversion this pullet was thrown on account of the influence of ancestral blood. Strange things are liable to happen under this law, yet it is true that like produces like.

4. Variation. The more closely related the purebred sire and dam and the closer the resemblance they have in type and color, the less variation there will be in the progeny; that is, the more closely the offspring will resemble the parents. Desirable characters have become fixed and the undesirable have been eliminated. This small amount of variation is because the blood lines are the same. This explains the advice sometimes given, that if a breed is to be improved or a new strain established there must be close inbreeding, that the dam should be bred to her own brother or her own sire if there is a close resemblance in points of excellence. Many breeders believe that this close inbreeding weakens the vitality and undermines the stamina of the progeny. On the other hand it is urged that such a probability is prevented by selecting for the sire and dam only individuals of marked vigor and vitality.

5. Persistent and diminishing characters. Whatever the mating may be, both sire and dam will have imperfections. These imperfections will show more or less in the progeny. If the desirable characters overbalance the imperfections the mating will be considered a success; and if, as the breeding continues, the imperfections gradually disappear and the good qualities are intensified, then the breeder knows that he will be able to establish a worthy strain. If on the other hand the imperfections grow stronger and the desirable characters diminish, it is evident that the line of breeding should be abandoned. A new mating should be tried. The tendency by which a character, on the one hand, persists and intensifies or, on the other hand, diminishes and disappears, we call the principle of "**persistent and diminishing characters.**" Its operation is quite hidden from the view of the breeder. Whether the desirable characters will become fixed and intensified in any line of breeding or whether they will diminish and disappear depends on the blood lines represented in the sire and dam, upon the discordant elements that may have been introduced in the line of ancestry and have not been bred out, and upon the prepotency of the mates.

6. **Transmission.** In sexual reproduction the male determines color, fecundity and prepotency, while the female determines type, size, temperament, and vitality. If these statements are true, it follows that the male should excel in color, prepotency and fecundity. It has been found that trap-nested hens with high egg records do not reproduce the quality of high production, but if a male bred from a hen with a high record for production is used in the breeding pen, the invariable result is an improvement of productiveness in all the offspring. If greater size is desired, large hens should be used for breeding.

7. **Intensifying defects.** Similar defects must not be present in both sire and dam, else they will be intensified and never bred out. If a defect occurs in the female it must be offset by an excellence in the male. The practice of offsetting the evil with the good should be practiced on every farm where pen mating is used. It will work a marked improvement in the flock.

8. Physical deformities in reproduction are the result of physical defects and occasionally may be due to injury or to malnutrition or to faulty incubation.

We conclude, therefore, that the very essence of poultry breeding consists in the selection of variations which appeal to the breeder and so intensifying them by repetition that they become fixed characters.

We have found that breeding is an artificial process, yet scientific. By it man seeks to mold and develop groups of individuals for the attainment of certain desired ends. There are several varieties of breeding which will be described in detail on the following pages.

Varieties of Breeding

1. Mongrel breeding. This occurs when several varieties are thrown together and **interbreeding** is followed from year to year with little attention paid to selecting, culling and mating. Often the best are selected for market, leaving the undesirables to perpetuate the flock. This practice is fast disappearing from the farms, and the rule now is to find purebred flocks scattered through the farming districts in every county.

2. Cross-breeding consists in breeding together two of the standard breeds or varieties. This is often done under the plea that crossing the two breeds increases size, vigor and productiveness. Occasionally some real benefit may appear but, if followed up indiscriminately, it will result in deterioration, and mongrelism will follow. However, cross-breeding is often used to advantage in producing new breeds and in breeding up a flock of poor quality to standard excellence. In this case new males of the same breed are purchased each year and, if this is followed up, a mongrel flock may be brought up to show-bird excellence. There is nothing to be condemned in this practice as it will result in better flocks than would be produced by mere mongrel breeding.

3. In-breeding. When full brothers and sisters are bred together we call it in-breeding, and, if this is followed up from year to year, we call it in-and-in-breeding, or close in-breeding. In-breeding as described results in deterioration and is to be avoided as far as possible. If this system is used judiciously, selecting only strong and vigorous mates, it may be used in the hands of the expert to establish racial characters and breed out imperfections. We do not advise it for the average breeder. We know what happens when close relations intermarry in the human race. Every farmer knows from experience what happens to the herd of hogs or other farm animals if brothers and sisters are bred together year after year. There is deterioration in size and vigor, and the herd becomes unprofitable. Line breeding, however, as will be shown in a subsequent section, can be practiced by the farmer, even though his equipment is meager.

4. Out-crossing. This consists in introducing new and unrelated blood of the same variety into a pen or flock. This is a very common practice on the farm and occurs when new breeding cockerels of his breed are purchased from year to year. Out-crossing avoids the degeneration which sometimes accompanies in-and-in-breeding but is attended with more or less risk. The new blood may contain some hereditary taint, entailing weakness or disease, or may lack in prepotency or represent an unproductive strain or possess blood which does not blend with the breeder's strain, thus destroying in a single season the work of long years of careful breeding. Out-crossing should not be unconditionally condemned, but the breeder should study the needs of his flock and fortify himself by a knowledge of the requirements of a good breeding male and purchase accordingly, and he will not be disappointed.

It is usually safer to purchase females to introduce new blood than to undertake to do it through the male. The male is more than half the flock and, if an error is made in his purchase, the whole flock is injured. If an outstanding female is purchased and her blood blends with the breeder's strain, then her cockerels can be used to supply new blood for the whole flock.

5. Line-breeding. This is a system of in-breeding by which vigor, shape, color and productive power are maintained. By this system of breeding, size and vigor are often increased, there is no loss of prepotency or stamina, and desirable characters are established and maintained. Line breeding is begun by selecting foundation stock as near the ideal as it is possible to obtain. The breeder must study carefully the requirements of the Standard of Perfection, so that he may know all the disqualifications as well as the points of excellence of the breed.

The picture that is made in his own mind after diligent study, is the ideal toward which he must strive. The ideal having been formed, he must select a female that has few defects and that possesses the shape, color, carriage and quality that measure up to his ideal. If he decides to have more than one female in the pen, they should be as closely related as possible and should harmonize with the ideal. Each hen should be leg-banded, and all should be trap-nested, and the eggs from each dam numbered. When the chicks hatch they should be toe-marked or wing-tagged, so that the progeny of each hen can be identified when matured.

The male should also conform to the ideal as closely as possible, and he should harmonize with the females. If there are any defects in the females they should be offset by corresponding points of excellence in the male, and vice versa. If the male selected is a cock, the females should be pullets; if the females are hens, the male should be a cockerel. Now it may happen that when more than one female is used it will be found that the progeny of one of the females is far superior to that produced by the others. In that event, her offspring should be chosen to continue the breeding line. The accompanying chart shows just how the mating can be made from year to year. It shows how a male line of breeding can

Modern type half-monitor house with open front

be established in which the blood of the male will predominate. Also a female line is established in which the blood of the original female predominates, and this is done without close inbreeding. In the third and fifth year the two lines can be brought together in such a way that the sire and dam used will not be closely related. The system advised is to use a sire unrelated to the dams. If this is done, and the blood of the sire harmonizes with that of the dam, then the breeding lines can be established and carried on without the necessity of using any close relationship in any of the matings.

The points to be emphasized in selecting the foundation stock are vigor, type, color, productiveness and prepotency.

A study of the chart will explain how these characters can be maintained without loss.

Cautions

If brother and sister are mated they should be of pronounced vigor, vitality and prepotency.

If the first mating in line-breeding produces inferior stock, the mating should be abandoned and a new one tried.

If occasionally a sport is thrown, the breeder should not be discouraged. By selecting the best each year this tendency will be overcome.

If new blood is to be introduced into a strain it will be safer to do it through one or two choice females than through a male.

If the original sire and dam are not closely related, there need be no close inbreeding. If they are closely related, how can two bloodlines be made out of one?

CHART OF LINE BREEDING

First Year

Ckl. (A) — Hen (B) → (C)
C1=Males=1/2 A+1/2 B
C2=Females=1/2 A+1/2 B

Male Line. — **Second Year** — **Female Line.**

Coo. (A) — Pullet (C2) → (D)
D1=Males=3/4 A+1/4 B
D2=Fem.=3/4 A+1/4 B

Hen (B) — Ckl. (C1) → (E)
E1=Males=1/4 A+3/4 B
E2=Fem.=1/4 A+3/4 B

Third Year

Cook (C1) — Pullet (C2) → (F)
F1=Males=5/8 A+3/8 B
F2=Females=5/8 A+3/8 B

Ckl. (E1) — Hen (C2) → (G)
G1=Males=3/8 A+5/8 B
G2=Fem.=3/8 A+5/8 B

Ckl. (D1) — Pullet (E2) → (H)
H1=Males=1/2 A+1/2 B
H2=Fem.=1/2 A+1/2 B
(H) ← (E1) Ckl — (D2) Pullet

Fourth Year

Cook (D1) — Pullet (F2) → (I)
I=Males=11/16 A+5/16 B
I2=Fem.=11/16 A+5/16 B

Ckl. (G1) — Hen (E2) → (J)
J1=Males=5/16 A+11/16 B
J2=Fem.=5/16 A+11/16 B

Ckl. (F1) — Hen (D2) → (K)
K1=Males=11/16 A+5/16 B
K2=Fem.=11/16 A+5/16 B

Cook (E1) — Pull. (G2) → (L)
L1=Males=5/16 A+11/16 B
L2=Fem.=5/16 A+11/16 B

Cook (D1) — Pull (H2) → (M)
M1=Males=5/8 A+3/8 B
M2=Fem.=5/8 A+3/8 B

Cook (E1) — Pull. (H2) → (N)
N1=Males=3/8 A+5/8 B
N2=Fem.=3/8 A+5/8 B

Fifth Year

Cook (F1) — Pull. (I2) → (O)
O1=Males=21/32 A+11/32 B
O2=Fem.=21/32 A+11/32 B

Cook (G1) — Pull. (J2) → (P)
P1=Males=11/32 A+21/32 B
P2=Fem.=11/32 A+21/32 B

Ckl. (K1) — Hen (H2) → (R)
R1=Males=19/32 A+13/32 B
R2=Fem.=19/32 A+13/32 B

Ckl. (J1) — Hen (H2) → (S)
S1=Males=13/32 A+19/32 B
S2=Fem.=13/32 A+19/32 B

Ckl. (K1) — Pull. (L2) → (T)
T1=Males=1/2 A+1/2 B
T2=Fem.=1/2 A+1/2 B
(T) ← (I1) Ckl. — (J2) Pullet

Explanation of Chart

A little study of the chart shows that there are three lines of matings. The first in which the blood of the original male predominates; another in which the blood of the original female predominates; and a third in which the mating results in offspring in which the blood of male is equal to that of the original dam.

An outstanding cockerel having been found, he is mated to a hen or hens of equal quality. The cockerel is unrelated to the hens and is known as "A" and the hens as "B." The progeny comprise the group "C" of which half are cockerels, "C1," and the remainder are pullets, "C2." If the mating gives good results the line of breeding is continued.

In the second year "A" is mated to his daughters "C2," and the offspring from this mating comprise the group "D," and each individual in this group represents three fourths the blood of "A" and one fourth the blood of "B."

In the same year another pen contains a cockerel "C1" from the first mating and he is mated to his dam "B." The result of this mating is group "E" consisting of cockerels "E1" and pullets "E2," each individual containing one-fourth the blood of "A" and three-fourths the blood of "B."

In the third year similar matings are made, a cock being used with pullets and a cockerel with hens. In this year also the male and female lines are brought together by mating a cockerel from the male line "D1" with a pullet from the female line "E2." The result of this mating is one-half the blood of "A" and one-half the blood of "B," exactly the same proportion that we had from the original mating. Again in the fifth year the two lines can be brought together with the same result. When a cockerel is mated with a pullet, both are supposed to be of pronounced vigor and prepotency, early hatched, and widely separated in blood lines. It will be noted that in all the other matings shown in the chart, brother and sister are not mated together and a cock is always mated with a pullet and a cockerel with a hen.

By following this system there is not at any time any very close relationships in the matings. Vigor can be maintained and by selecting only the best each year good qualities can be preserved and intensified.

Many other combinations besides those named can be used. If the original male has unusual vitality and is a prepotent breeder he can be used for three successive years and this will intensify his blood in the male line. So also can the females be used for three years intensifying the blood of the original dam in the female line. Probably for all practical purposes the farmer will not wish to use his males and females longer than for two seasons.

Dinner time in the poultry yard

Chapter IX

Mating and Culling

THE following are important steps in breeding:

The Ideal. The first step in breeding is determining the ideal. This is formed by a study of the breed as exemplified in living birds or by a careful study of distinguishing characters as set forth in the Standard of Perfection. The ideal is necessary to progress. Otherwise all efforts to produce or improve a breed will be aimless. Chance will be the controlling factor. When the mind has a perfect picture of the type and color of the variety chosen, the breeder is then ready to select his foundation stock with discrimination.

Selecting the Foundation Stock

Success or failure will depend on this step. Here are some of the general principles concerned in the selection of breeding stock:

1. Sire and dam should not be related closely if it can be avoided.

2. Reject all specimens from the breeding pen that have ever been sick or that have any constitutional weakness.

3. Insist that every specimen shall approach the ideal, as near as possible, in type, color, and stamina.

4. Avoid all disqualifications in either sex. The more important general disqualifications are the following: Side sprigs on single combs; absence of spike on rose combs; feathers on shanks or toes of clean-legged varieties; absence of feathers on toes or shanks where they should occur; deformed beaks, wry tails, crooked backs, crooked keels, squirrel tail; color of feet other than required by the standard; color of face and ear lobes white, when the standard requirement is red, and vice versa.

4. Avoid birds too large or too small for the breed. In most cases the standard weight will give best results.

How to Select the Male

In selecting the head of the flock or pen, the following factors should be considered: Vigor, vitality, prepotency, productive power.

Vigor. Physical stamina and energy of action are the proof of vigor. If I were asked to select a male of strong constitutional vigor, I would insist on the following: Legs well set apart; shanks strong and not too long, back broad at shoulders, breast full, body deep and wide and the span between the lower point of keel and ends of pubic bones at least

A knock-kneed rooster, showing low vigor and vitality. Do not select males like this for breeding.

three fingers, eye full and bright, head wide between eyes, and a bird that has a good appetite and is pugnacious and courageous.

Vitality. There is a difference between vitality and vigor. Vitality is the grip on life. It is that invisible something that inheres in the individual. A fowl may be strong and vigorous but lack in vitality because it lacks resistance to unfavorable influences. When a bird has good vitality it resists disease and survives while others quickly succumb. A male bird that survives two or three seasons in the breeding pen certainly has good vitality.

Prepotency. This is the power that an individual has to perpetuate its qualities. Vigor, vitality and prepotency often go together, but a bird may have the first two qualities but lack in prepotency. When a male stamps his individuality upon his progeny he is said to be prepotent. Such a male is invaluable. Prepotency is indicated by the jealous attention of the male to the female. Strongly fertile eggs are an evidence of prepotency in the breeding stock.

A sure method of detecting prepotency is the trap nest. Walter Hogan in his "Call of the Hen" gives a method which is worth considering. A point, known as "A", is found where the skull joins the atlas along the median line over top of head; and a similar point, known as "B", is found on the side just behind the ear where the skull joins the neck. If A is behind B, prepotency is strong; if A and B are opposite or in the same plane prepotency is only average; but if B is behind A prepotency is weak. Some consider prepotency a "breeders' superstition" and that the Hogan test is only a "myth." By many breeders this test is considered reliable and that it should always be used in selecting special matings. It is worth a trial.

An individual may be prepotent in one character but not in others. Again, an individual may be prepotent in several characters, such as type, fecundity, vitality and color. When a male is found to be prepotent he should be greatly treasured on account of his influence upon the whole flock.

Producing power. It is quite generally recognized that high egg production does not always come from females whose dams were good producers. The male is an important factor in securing large egg production. If the male is derived from a hen having a high record for egg production, his progeny will be good producers. The trap nest is necessary to select the high producing hens. In addition to his breeding the male should pass the culling tests usually applied to the female. The pubic bones should be straight, not unduly thickened at the ends, and should be separated by at least one finger. The capacity measure should be at least three fingers, this being determined by the span between the lower point of keel and the ends of the pubic bones. The tail should be well developed and the comb, wattles and earlobes should be fine in texture.

How to Select the Female

The qualifications of the female are practically the same as with the male. The head points, however, are finer and the cranium is narrower. Special emphasis should be given to type and size. A hen with a baggy abdomen should not be used as a breeder. A hen that produces abnormal eggs should be removed from the breeding pen. A hen with any bad habit, such as egg-eating, feather pulling or laziness should be cured or rejected.

Selecting and Mating by the Trap Nest

There is no surer method of determining laying capacity than by the trap nest. While this method is not practical on many farms yet the farmer can profit by the discoveries made through this means in the experiment stations and egg-laying contests. The valuable data collected by investigators through trap-nest records have given a great impetus to the study of poultry problems. Two investigators, widely separated in space but widely known in poultry circles, are Professor James Dryden of the Oregon University Experiment Station and Dr. Raymond Pearl of the Maine University Experiment Station. Their deductions from trap-nest experiments have thrown a world of light upon the problems of fecundity and heredity. We take pleasure in quoting from these authorities, believing that the facts which they present should have wide circulation.

Professor Dryden's Conclusions

"Regardless of any question of prepotency, the selection of breeding stock on the basis of production record is a certain method of increasing production.

"Some individuals have greater power of transmitting high fecundity than others of the same breeding.

"Good layers are not always produced by good layers, nor are poor layers always produced by poor layers.

"Rapid progress can be made by the breeder if he tests the breeding quality of his stock by using for breeding those hens and males whose progeny has shown high production.

"Rate of laying is within certain limits and accurate measure of egg-laying capacity.

"Good laying capacity is indicated by heavy production in any two months of the year.

"Late laying in the fall does not always indicate good layers."

Dr. Pearl's Conclusions

"One of the least understood phenomena in genetics is prepotency. It is customary to regard an animal as prepotent in breeding for performance when the progeny of that individual uniformly tend to resemble it closely in respect to the character bred for, regardless of the other parent in each mating.

"High fecundity is not inherited by the daughters from their dam.

"High fecundity may be inherited by the daughters from their sire independent of the dam.

"A low degree of fecundity may be inherited by the daughters from either sire or dam or both.

"Winter egg production is the best available measure of innate capacity in respect to fecundity.

"Variations in fecundity are not determined by the number of visible oöcytes on the ovary. This number varies by actual count from 900 to 3,600.

"Breeding for high fecundity requires the use of such females only as have shown themselves to be high producers, since it is only from such females that there can be any hope of getting males capable of transmitting high laying qualities; and the use as breeders of such males only as are known to be sons of high producing dams, since only from such males can we expect to get high producing daughters."

Conditioning the Breeders

Selecting foundation stock is important, but to stop at this point is to court disappointment. The stock must be conditioned for best results just as a piece of machinery must be kept oiled and in trim condition to give good service. That breeding stock should receive different treatment from stock that is being forced for egg production is generally admitted. Condiments, high stimulants, and hot mashes should be

Mr. Osburn determining the thickness of the pubic-bones

avoided. A balanced ration, pure water, plenty of exercise, abundance of sunshine and oxygen and extreme cleanliness are all indicated.

Culling the Flock

Culling is the final step in poultry breeding. In nearly every farm flock there are three classes of undesirables—the weaklings, the defectives and the non-producers. The object of culling is to weed out the weaklings, to swat the slackers, and to discard the defectives. When the work of culling is completed and all the culls have been conditioned and shipped to market, there should remain but two classes, the breeding flock and the laying flock. Many fowls that would not pass muster as breeders can be used in the laying flock.

Culling should be made at least four times in a year—March, June, September and December. In December breeding pens are made up. The defectives are put with the laying flock, and those that meet standard requirements are saved for the breeding pen. At this culling some weaklings will be found that should not be saved for egg-production. These can be conditioned for market along with any surplus cockerels that are not suited to be sold as breeders. In March, all the hens that will prove unprofitable producers can be detected, for at this season every hen that is worth saving will lay.

In June, another culling of the layers will weed out a few that have passed the productive age and some poor producers that were overlooked in earlier cullings. Early hatched cockerels can also be culled at this time, some to be caponized, others to be marketed as broilers.

In September a thorough culling should be made. It is possible then to detect the early molters, and the condition of plumage and pigmentation makes it quite easy to pick out the good layers. At this time, also, the young stock can be selected for the fattening crate and the laying and breeding pens.

As an aid to memory in making culling demonstrations I have used the word "T-e-a-c-h", each letter in this word representing an important step in the work: Type, Energy, Anatomy, Condition, Habit.

Type is conformity to an ideal in shape. In culling non-layers, type is a very important item. We have discussed the different types—meat type, dual purpose type, and egg type. What do we mean by an egg type? Evidently that conformation of body and bodily structures which is conducive to highest egg-production. Is there an egg type and, if so, what is it? Breeders are not agreed upon any single type of fowl as better adapted to egg-production than all others. Some prefer the wedge-shaped. Such a fowl is full in the abdomen but the body gradually tapers toward the neck. If the hen is viewed from above or the side or the rear it presents the form of a wedge. The wedge-shape appeals to them because it indicates large abdominal capacity.

Capacity is shown by the space between the keel and pubic-bones

The slab-sided type is advocated by some. The body presents the appearance of a flat side-surface which is deep and long. Such hens are usually narrow in body, but they are counted as good layers. Another egg-type advocated by some is the spherical body. Viewed in any direction the outline of the body is round. The Wyandotte is a good illustration. Some Wyandotte breeders have had courage to break away from standard requirements and are now breeding for a longer body with a full breast, believing that type to mean larger capacity and greater egg-production. Still another type that has a large number of advocates is the oblong type. The body is long and wide across the back, and of good depth. The breast is full, the keel long, and the abdomen full. Viewed from above or on the side the shape is oblong. This type is illustrated in the White Leghorn and the Rhode Island Red, as well as many other breeds. This type probably has more friends than any other.

Type is also indicated by the thickness and conformation of the public bones. If they are one-eighth of an inch or less in thickness, the egg-type is indicated. If they are moderately thick, one-fourth of an inch or a little more or less, the dual purpose type is indicated. If they are very thick, one-half inch or more, the meat type is suggested, and such hens tend to lay on fat rather than produce eggs. In making a study of the dual purpose breeds it will be found that moderately thick pubic bones are compatible with good egg-production.

Egg type is also indicated by the head points. A bright eye, a short beak, a large comb, fine in texture, large wattles, and a face with a close fitting skin are indicative of the egg type. If a hen falls short in these requirements she may be suspected of being unworthy of a place in the flock.

Energy. Energy is the power to do work. This power depends upon health, vitality, prepotency, capacity, and temperament.

Health is a condition in which the functions of the body are performed without friction; disease is an abnormal condition. Only healthy fowls should be allowed to remain in the flock. If diseased, they will soon cease laying, and, if they lay, their eggs will be unfit for human consumption.

Vitality is essential to production. If weak in vitality the hen should be rejected. Sometimes, however, a hen may naturally have a good grip on life, but its vitality has been weakened by external or internal parasites. If the parasites are destroyed, she will put on new life and become productive.

The value of prepotency has been discussed. Hens that lack in prepotency should be culled from the breeding flock, but may be retained in the laying flock.

Capacity is an index of energy. By capacity is meant an abundance of room for the operation of all the machinery engaged in egg-production. A machine in cramped quarters will turn out only a limited amount of finished product because it cannot handle the raw material in abundance. If the digestive and reproductive organs which constitute the egg factory are in a cramped abdominal cavity they cannot be expected to yield a large output in eggs. Capacity is usually determined by the span between the ends of the pubic bones and the lower point of the keel. This measurement should be three fingers or more. If the measurement is four or five fingers, capacity is large and the hen is probably a good producer. If less than three fingers there is reason to doubt the quality of the hen.

This caution should be given. There are times when a good producer will be found more or less contracted in the abdominal region. This may happen at the close of a long siege of sitting or during the period of the molt. This must be taken into account in culling for this defect. Again, capacity may sometimes be abnormally large, as with a hen excessively fat or with a drooping abdomen. A baggy abdomen, which hangs below the keel, indicates a breaking down of the egg organs, and such a hen should be marketed at an early date.

Temperament has something to do with the energy of a fowl. Temperaments are described as nervous, sanguine, lymphatic and bilious. A nervous temperament means greater energy and more work accomplished than could be possible with a bilious temperament. The Buff Cochin has a bilious temperament and produces few eggs; a Plymouth Rock has a sanguine temperament and is a good average layer. A White Leghorn has a nervous temperament and excels in egg-production. Many individuals of the dual purpose breeds are nervous in temperament and excel in production. This is illustrated in the White Rock hen, Lady Show-you, the champion in the first Mountain Grove, Mo. contest. She was nervous, exceedingly active and energetic, always searching for food. Temperament is a good guide in selecting the good producers of a flock.

Anatomy. This term applies to structural characters. The most important structural characters which have a bearing on egg-production are the **pubic bones.** In a good layer these are thin, straight and well spread apart.' The spread is generally measured by the tips of the fingers. A finger measures about one-half inch to three-fourths of an inch. When a hen is in laying condition the distance between the pubic bones is three fingers to five fingers in a good layer. When the measurement is only two fingers or less the test is considered unfavorable. Nearly all hens during the molt, or while sitting, will have a much smaller spread than while laying, and this must be considered in culling. Watch for abnormalities. There are certain anatomical defects which should always be sought in culling demonstrations and, if found, the hen should be rejected. An important one is crooked pubic bones. If these are much curved so as to approach each other at their points and are tied together with rigid skin and muscle, the hen is a poor producer. Such a condition interferes with the process of laying, so that it is attended with more or less pain. Hens with this defect lay fewer eggs. Undoubtedly crooked pubic bones are the cause of blood-stained eggs. Such hens should not be used as breeders, and they are unprofitable layers.

Another defect is the crooked breast bone. This defect is hereditary, and birds affected should be discarded. A deformed beak and a crooked back are other defects that count against the laying performance of a hen.

Occasionally hens are poor layers on account of structural defects in the egg organs. Usually these cannot be detected during the life of the hen, but sometimes the shape of the egg, the texture of the shell, or the condition of the contents are tell-tales which indicate that the hen should be removed, at least, from the breeding pen.

Condition. A very good indication of the laying performance of a hen is her condition at different seasons. In this connection we should consider the condition of plumage, skin, flesh and pigment. In September, after a long period of laying, the plumage is worn and faded. This indicates a good layer. The skin of a good layer upon the abdomen is found to be soft, velvety, and pliable. A good layer, if fed properly, carries a good supply of flesh. This is noted by the condition of the keel, which is well covered. A poor layer is abnormally fat or unusually thin, with little flesh on keel. At certain seasons pigmentation is a very good clue by which to detect the poor layers. In yellow skinned breeds the vent will continue yellow in the non-layers, but becomes pale or pinkish-white among the good layers. The color disappears from the beak, skin, eye, and legs of the good layer, but a poor layer will retain the yellow beak and skin and brilliant plumage.

Habit. This is the final test of productiveness. Habit is the method of life. Three habits may be mentioned as bearing on egg-production, viz., molting, feeding and exercising.

Molting, or shedding of feathers, is a common habit of poultry, but the time of molting is not fixed. Late molters and quick molters are invariably good layers. The reason assigned is that late molting lengthens out the laying season, while quick molting shortens the resting period. Molting and egg-production are not compatible. When a hen begins to molt she slackens in egg-production.

The hen that retains her pigmentation and molts in early summer is counted as a poor layer and should be culled from the flock. In culling on the basis of an early molt, this caution should be observed, that good layers are sometimes thrown into an early molt because of environment or a sudden change in feed. Before such hens are thrown out they should be tested along other lines for further proof of unproductiveness.

The habit of sitting is very trying among some breeds. The persistent sitter, the one that, when broken, lays only a small clutch of eggs before becoming broody again, is usually an unprofitable layer.

The habit of feeding varies with different hens. The poor layer is indifferent. She has no occasion to hustle for food as she is not a producer, and for this reason she is early to roost and late in leaving the roost.

The habit of exercise characterizes the good layer. She is a good scratcher, and scratching makes short claws. She is a hustler and works till sundown and is up before sunrise.

Score Card for Culling Demonstrations

1. Type	20	Body type	5
		Thickness of Pubic Bones	10
		Head Points	5
2. Energy	20	Vitality and vigor	5
		Prepotency and Temperament	5
		Capacity	10
3. Anatomy	20	Spread of Pubic Bones	10
		Shape of Pubic Bones	5
		Structural Defects	5
4. Condition	20	Plumage	5
		Pigmentation	5
		Skin	5
		Flesh	5
5. Habit	20	Molting	10
		Feeding	5
		Exercise	5
	100		100

Trap nests makes it possible to keep a laying record of each bird

No bird should be retained in the flock that scores less than 75 points. Example: A hen that should be laying in July is found to have pubic bones one-half inch thick—deduct 10 points; the capacity measurement is only two fingers—deduct 10 points; the keel is crooked—deduct five points; the pigmentation is yellow in beak, skin and legs—deduct five points; the hen is in full molt and there is no evidence that she has been laying—deduct 10 points. The total deductions amount to 40 points. Her score therefore is 60 points. Ordinarily a hen with as many cuts as here given would be found defective in several other tests. In this particular case we assume that the other points are good. A hen should not be rejected on one or two cuts unless they are very decidedly unfavorable. The practice of throwing out hens simply because the pubic spread is small may lead to serious error. The bird should be tested in all points and the decision made according to the preponderance of evidence.

Mating the Farm Flock

On the average farm there is little or no effort made towards line breeding. There are not even separate pens in which to keep the breeders apart from the laying stock. The stock is given free range, summer and winter. Some little effort is made toward mating the flock by culling out everything that should go to market and by purchasing new cockerels, so as to maintain the vigor of the flock. A few words along the line of mating the general flock may be helpful. It would be far better if a few breeding pens could be made so

A self-feeder for mash, charcoal and oyster shell

as to keep the breeding stock separate from the layers. Then the laying stock could be forced for winter eggs, while the breeders could be given the care and feeding required to produce fertile eggs that would hatch sturdy chicks. Where this cannot be done, the following course is advised:

1. Cull out all the undesirable females.
2. Remove from the flock all late hatched cockerels. If brothers and sisters are to be mated they should be of outstanding quality and of pronounced vigor and stamina.
3. Keep the yearling cock birds, if they have proven of good breeding quality.

4. If a pure-bred flock, select males strong in color and females strong in shape and size.

5. Mate so that the defects in one sex will be counteracted by strong points in the opposite sex.

6. Avoid all disqualifications in either sex.

7. Select males and females that qualify under the tests for the type desired, whether meat, dual purpose, or egg type.

8. Use great care in introducing new blood. It means the introduction of new characters which may not harmonize with your own strain; further, it may mean the introduction of some weakness or latent disease that will bring disaster.

9. Avoid females that produce abnormal eggs, such as undersized, oversized, elongated, round, rough-shelled, or uneven eggs.

How to Mark the Breeders

It will be found helpful to mark all the breeding stock. Where there is a general flock, the pullets and cockerels can be leg-banded each fall, a different color being used each year. The celluloid rings are good for this purpose. If this is done. there will be no difficulty in distinguishing pullets from hens. Where line breeding or pen breeding is pursued, numbered leg bands or wing markers should be used. The birds of each pen can be given bands with a separate color, and different colors can be used for each year, if necessary, to distinguish pullets from one-year-old or two-year-old hens.

Chapter X

The Poultry House

A SUCCESSFUL poultry house combines the following characteristics:

1. **It is dry.** This is secured by having a concrete floor laid upon cinders, broken stone, or coarse gravel, and elevated about one foot above the surrounding ground. The roof and walls are water proof and a good circulation prevents the condensation of moisture from the fowls upon any portion of the room.

2. **It is well ventilated.** This is accomplished by windows. shutters, screens, or ventilators placed on one side of the building so as to prevent drafts. The top, two ends and rear side of a poultry house should always be air tight. The ventilation, therefore, should be secured from the front. The shutter method is probably the best as it prevents the entrance of rain and snow and can be kept open in all seasons. A frame covered with fine wire netting will be found very satisfactory. It has the advantage of admitting light as well as pure air. The glass windows should be adjusted so that they can be opened when weather conditions require.

By this system ventilation is accomplished by diffusion. There is a constant interchange between the warm air within and the pure air on the outside. Such a poultry house will never be stuffy or ill-smelling. Other methods of ventilation are by means of ventilators placed in the roof, by canvas-covered doors, and by an open front. Ventilators in the roof often serve a good purpose; the pores of canvas soon become filled with dust so as to prevent the fresh air from sifting into the house, and canvas shuts out the light; the open front is very popular and will be satisfactory if provision is made to protect against driving storms.

3. **It is well lighted.** Windows should extend from upper to lower plate so as to admit the sunlight to all portions of the floor. Sunshine is a germicide, as well as oxygen, and it is not possible to have the supply too abundant. Windows should be

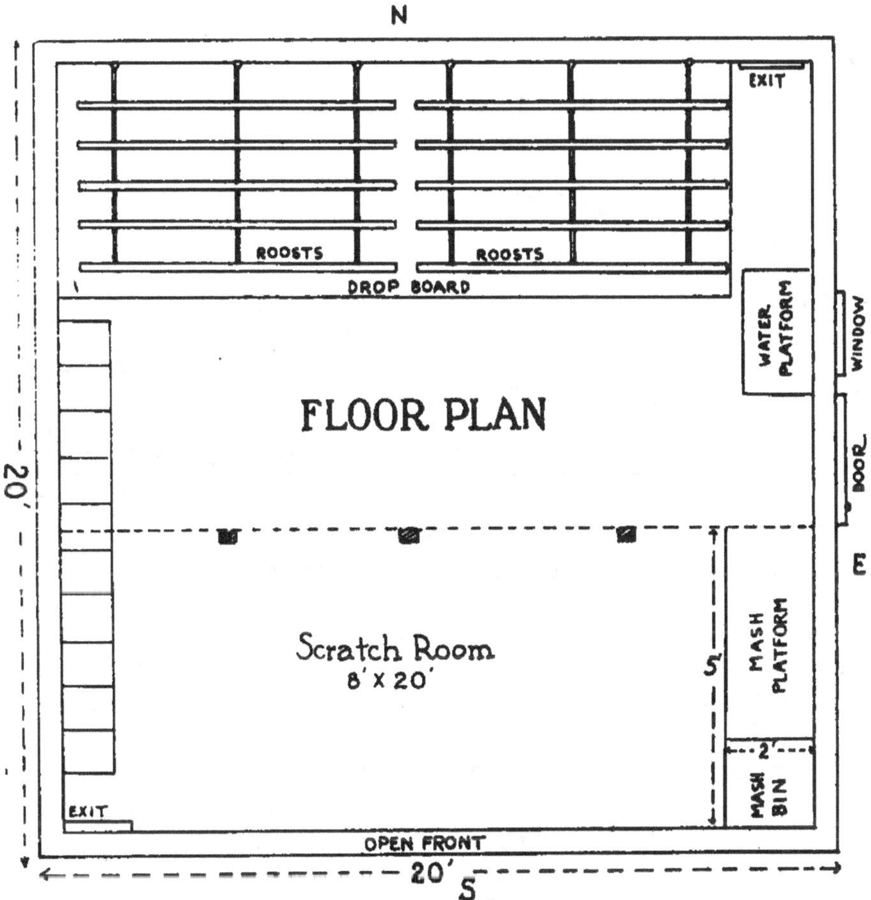

Floor plan of ideal farm poultry house for 100 laying hens. See elevation plan on next page. This house can be built for $150 to $200

covered on the inside with half-inch wire netting to prevent the breaking of lights by unruly fowls.

4. **It is comfortable,** warm in winter and cool in summer. If the shed roof type of house is used, the slope of the roof will be toward the north and hence will be protected from the direct rays of the sun. Such a house will be much cooler in summer than where the roof has a large exposure to the south.

Such a house will also be comfortable in winter, as it will get the benefit of the sun's rays. If necessary, canvas curtains should be provided on rollers to be pulled down as a protection in severely cold weather. Houses should never be heated artificially in winter as the fowls will thrive better with a reasonable amount of cold.

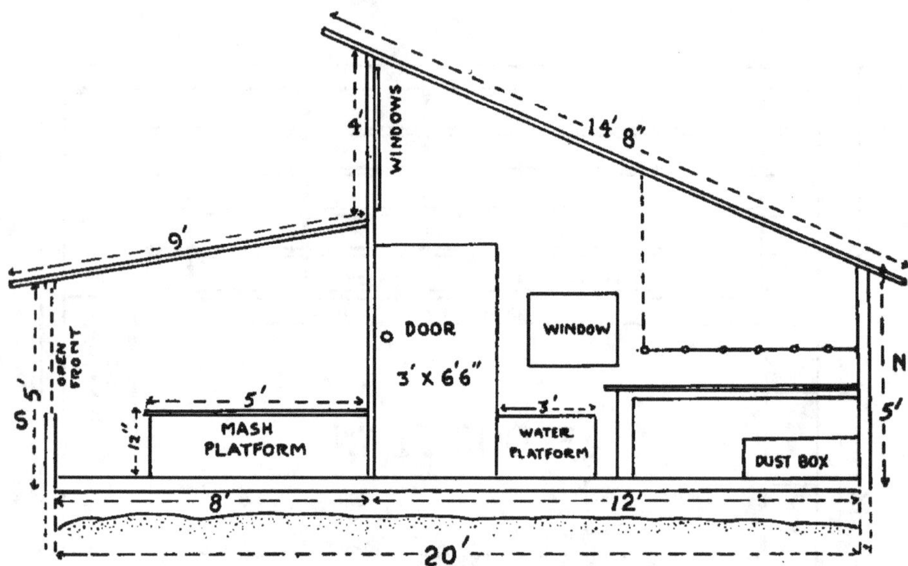

End elevation of ideal farm poultry house. See cut of complete house on next page

5. It is roomy. The floor space is adequate for the number of fowls. Four square feet should be allowed for each hen. A house 20x20 feet will accommodate 100 hens. A house 20x40 feet will accommodate 200 hens. Overcrowding does not favor egg-production. It is better to provide too much room than too little. Sometimes a flock of ten hens in roomy quarters will produce more eggs than a hundred in crowded conditions, even though given like care otherwise.

6. It is convenient. The nests, the feeding platform, the water fount, and other devices are all so arranged as to save labor for the attendant. The house will be accessible so as to save labor in providing water, feed, and litter.

7. It is vermin proof. The floor is of concrete to prevent the entrance of rats, mink and other vermin. All nesting and roosting places for sparrows are closed, and all hiding and breeding places for mites and lice are avoided as far as possible. Nests and roosts are removable so that they can be taken outside and thoroughly cleaned and disinfected.

8. It is inexpensive. A hollow tile wall would be ideal, but ordinary tongue and groove siding will answer all purposes. Paper roofing will serve many years if shingles are too expensive. If it leaks from any cause, it should be quickly repaired as a protection to the flock and to the building.

Location

The selection of a site for the poultry house is of prime importance. This should be on the highest, best drained ground, not too far from the farm residence. Such a location may be exposed to driving winds and rain and snow, but it is easier to build windbreaks than to drain low ground. A sandy loam or porous soil with gravel subsoil is the most desirable for poultry, for it eliminates stagnant pools of water which generally prove a menace to the flock. A clay soil is to be shunned, if possible.

If a suitable soil cannot be obtained, a few loads of gravel or cinders around the building will help matters. A site adjacent to the orchard is ideal, for it furnishes a place of forage for the flock, the trees furnish shade in the heat of summer, and fowls contribute their help by destroying harmful insects. If the ground slopes in all directions from the house, good drainage is insured. If the building is on a side hill, some provision must be made for drainage on the high side or else the floor of the house should be raised so as to be above the surrounding ground.

This is the way the farm poultry house built from plans on preceding pages looks when completed

The semi-gable type of poultry house

Types of Poultry Houses

What type of a house is most suitable to the farm? The shed roof has many friends. The advantages of this type are several: It furnishes a large frontage by which sunlight can be admitted to all parts of the building; it is more easily constructed and is less expensive than other types; as the roof has a northern slope it will be more lasting than other roofs because protected from the direct rays of the sun. The shed roof has the disadvantage that it is easily torn away and destroyed by heavy winds. When the wind strikes a sloping roof, its force is resolved into two forces, one parallel to its surface which has no effect, and the other perpendicular to its surface which exerts a downward pressure and tends to hold the building in place. But when a high wind strikes the front side of a shed-roof type of building there is no downward pressure and, unless the building is carefully anchored, it will be torn away by the wind. I have known shed-roof buildings to be torn to pieces and scattered on a distant field while other buildings with roof pitching toward the wind have been unmoved.

The next most popular and serviceable type of poultry house is the semi-monitor. This is illustrated on page 125. This is a more expensive type to build as it has a broken roof. The upper windows allow the sunshine to reach the back portion of the building. The front portion can be used for a scratch room and, if an open front is desired, ventilation can be secured in that way. Such a building should be made deep, 16 ft. to 20 ft. The deeper the building, the more favorable to the flock and the better the results.

A third type is the semi-gable. This is in common use and has advantages over the two mentioned above.

How to Build the Poultry House

The following simple directions may assist in building a semi-monitor house 20 feet square. The trenches should be dug at least 18 inches deep as a protection against frost and vermin. The dirt is thrown to the outside to be afterwards graded against the foundation. The forms for the foundation should be made of twelve inch or two six inch boards. After the forms are placed the trench is filled with the concrete made of four parts gravel and one part cement.

As the cement is being placed strong bolts with heavy washers are set so that the top of each bolt will extend about an inch above the top of the foundation plate. These are to anchor the building. The concrete is brought flush with the tops of forms and made level and smooth. As the foundation is made six inches wide a 2" x 6" timber can be used for plates. These are now bolted to the foundation. The floor space is now filled with cinders, broken stone or coarse gravel to a depth of six inches. As these substances are porous and contain air spaces they prevent moisture coming up from below, thus insuring a dry floor at all times. We are now ready for the concrete. A layer two inches thick of the same quality as used in the foundation is placed upon the broken stone and made level to receive the top dressing. After the concrete has begun to set the top coat is applied consisting of two parts of coarse, gritty sand, and one part cement. This should be made one inch thick and should be troweled level. After the top coat has begun to set a skim coat of pure cement and water is applied. This is to be very thin and should be troweled until a smooth hard surface is formed.

A convenient poultry interior. The roosts swing up so that the droppings board can be easily cleaned. A better position for the nests would be on the side wall as this would save floor space, needed by the hens

Unless the walls are to be built of hollow tile, the posts can be made of 2x4 inch stuff. If the building is to be 20 feet deep the posts in front should be 4 feet 6 inches high and in the rear 5 feet. If it is to be 16 feet deep they should be 4 feet long in front and 4 feet in the rear. In the building 20 feet square the rafters for the front pitch can be made of 2x4 inch stuff 9 feet long. This allows for a 1 foot projection at the eaves; and for the rear pitch the rafters should be 2x4 inch stuff, 14 feet 8 inches long. The roof supporters, three in number, should be 4x4 inch posts placed 8 feet from the front.

The siding should be tongue and groove 1 inch material free from knot holes. The sheeting should be matched flooring if a paper roofing is to be used. In the front elevation above front roof there should be five windows as shown on page 125. At least one of these should be hinged to allow of opening for ventilation. The open front should be provided with frames covered with wire screen. The door to the building should be on the east side. West doors should be avoided as prevailing winds are from the west. Exits and entrances for the fowls may be placed under the ventilators and at one corner in the rear. The perches should be made of 2x3 inch stuff 9 feet long, rounded on the edges. They can be supported by ½ inch iron rods which pass through holes bored in each perch. These rods are fastened to the rear wall by means of large screw eyes which are hooked into a ring formed at the end of each rod. This forms a hinge so that the perches can be raised and lowered at will. Two supporting rods and five perches in each section would be sufficient for 100 fowls. The perches should be one foot apart and arranged so as to be level. They should be placed about eight inches above the droppings board. The droppings board is made of half inch matched lumber in sections so as to be easily removed. They can rest upon 2x4 inch timbers, one nailed to the rear wall and the other supported by legs in front. The platform should extend in front of perches about twelve inches to provide an alighting place for the fowls. These droppings boards should be thirty inches from the floor. Nests can be built on the side walls or on an elevated platform between the roosting room and the scratch room. They should not be built on the floor, for then soft shelled eggs and cracked eggs will be eaten by

Osburn's poultry house

An inexpensive colony house

the fowls, thus leading to the egg-eating habit and, by this method, floor space is used that is needed by the hens. Platforms should be provided for the self feeders and watering vessels. All covers for nests, self feeders and hoppers should be inclined or they will become roosting places for the fowls. A bin for holding dry mash can be placed under the ventilators or in one corner of the scratch room. It should be mouse proof, so should be lined with tin and have a tight fitting cover. It is customary with long houses to have a room at one end for the storage of feeds and other supplies.

In the plan proposed for this house, the top of plate is five inches above the cement floor. This gives room on the floor for a deep litter that will not interfere with the opening of doors. If the house is built in more than one section, a partition of half-inch material should be built between the sections. The door between the sections should be light, of half-inch material, and hung on swinging spring hinges. The front portion of the floor space is used as a scratch room, where the grain mixture is fed. The rear portion can also be used for that purpose if necessary.

Some poultry keepers advise a dust box to be partly filled with soil, sifted ashes, road dust, dustyne or other material. It can be placed under the drop boards, and surely the hens appreciate it. The good points of a house such as described are ventilation without a draft, an abundance of sunlight reaching all parts of the building, a dry floor, it is roomy and comfortable, and rats and sparrows cannot infest it.

Lots

It is a great help to have one or more lots connected with every section of the poultry house. If winter eggs are to be forthcoming, the flock should not be allowed to roam over the farm, but they should have access to the outdoor air at least a portion of nearly every day. This cannot be accomplished without chicken tight lots. If there is a lot in front and one in the rear the ground can be cultivated, and while the hens are using one lot, a crop of rape, oats or other green feed can be grown in the other. In this way the soil is kept sweet and disease is warded off. If only one lot can be provided it should be disinfected frequently by sprinkling the ground with a solution of copperas, one pound to 50 gallons of water. Slaked lime is also good for disinfection if scattered over the ground. Another good method is to spade the soil. This gives oxygen

Mr. Osburn uses this convenient nestbox arrangement

Raising baby chicks is easy with a brooder house like this

and sunshine a chance to penetrate, and they are the great germ killers.

Other Buildings

Besides the breeding and laying house, other buildings will be found necessary, especially if large production is expected.

Incubator House. Under average farm conditions the incubator can be set in the house in some room where an even temperature can be maintained. The cellar, also, makes a good location. If chicks are to be produced on a large scale an incubator house will be needed. It can be built of stone with thick walls or of hollow tile and at least half of the room should be below the ground level. If an upper story is made it can be used for storage of feeds and supplies. Provision must be made for drainage and ventilation. The incubator should not be near a stove, or in a draft, or subject to direct sunshine. If the incubators are to be run in cold weather provision must be made to keep the room temperature at about 60 degrees.

Brooder House. This is built after the same general plan as the laying house, the difference being in size. A long brooder house without partitions would not be successful on account of drafts which are always found in such a building. It would be better to have several smaller houses. A room or house, 8x12 feet, would accommodate 100 to 200 chicks, and we doubt

A dropping board makes it easy to keep the house clean

whether more than that number should be crowded under one hover. A concrete floor is necessary to keep out vermin, and ventilation without draft can be secured by muslin windows.

Platforms for water, mash, grit, etc., if made accessible to the chicks, will be a great help toward keeping these supplies clean.

About two inches of coarse sand on the concrete floor will prevent rheumatism and leg weakness.

A handy brood coop for hen and chicks

Colony houses. Small colony houses for the growing stock are of great value. A house 3x6 feet, with a shed roof, will accommodate thirty to forty chicks. The floor should be tight and all openings covered with wire screen to keep out vermin. Two large doors in front may be provided, one of glass for light and the other covered with wire netting and muslin for ventilation and protection. Such a house should face the east as a protection against prevailing winds and should be elevated upon runners, stilts or stone piers as a protection against vermin. The runway into the house should be closed on the sides to assist the chicks in finding their way inside.

Brood coops. Galvanized iron coops are considered sanitary but are intolerable under a hot sun. Shed roof coops, made at home, answer well. They should have a floor bottom and a runway in front. A sliding door will protect against enemies at night. If closed at night holes in the side for ventilation should be provided.

The hen hatchery. Where chicks are to be hatched by hens some provision must be made for the sitting hens. A small house with compartments or a room set apart for that purpose is all that is necessary.

The feed room. One of the most difficult problems on the farm is to provide a room for grains and mill products required for feeding poultry and other farm animals that is absolutely rat- and mouse-proof. These vermin contaminate the food and bring disease and must be outwitted. The floors of bins should be covered with tin or iron unless there is a concrete floor and all windows and openings covered with wire netting. Mashes left in sacks become a harbor for vermin. It would be better to mix the mashes as soon as the materials are obtained and put them in vermin proof bins. Many farms are provided with machinery to grind grains so that cracked corn, corn meal, ground oats, etc. can be produced on the farm. Wheat can be taken to the local mill and bran, middlings, etc. obtained in that way.

Other buildings may be required, all depending upon how extended are the operations of the producer. Conditioning rooms or houses, storage places, and a detention house are of this character.

Cautions

Do not build the poultry house on low ground. Dampness breeds disease and disease spells loss.

Do not face the poultry house toward the north or the west. Prevailing storms are from those directions.

Do not put doors or windows on the west side of the poultry house.

Another convenient brood coop

Do not put nests under drop-boards upon the floor. It means vermin, egg-eating hens, and loss of floor space.

Do not close the front of building with glass. It means a damp building.

Do not set watering and feeding vessels on the floor, but on elevated platforms.

Do not allow leaks in the roof.

Do not permit cracks, knot holes or other openings on the north, east, or west.

Major Equipment

The major equipment comprises incubators, brooders, self-feeders, grain spouters, egg cabinets, trap nests and all other supplies of a durable nature. Most of this equipment can be purchased of supply houses at less expense than if made by the poultry keeper. Some, however, such as nests, self-feeders, and outdoor brooders (fireless) can be made at home with considerable saving of expense. Illustrations are given showing how some of these are made.

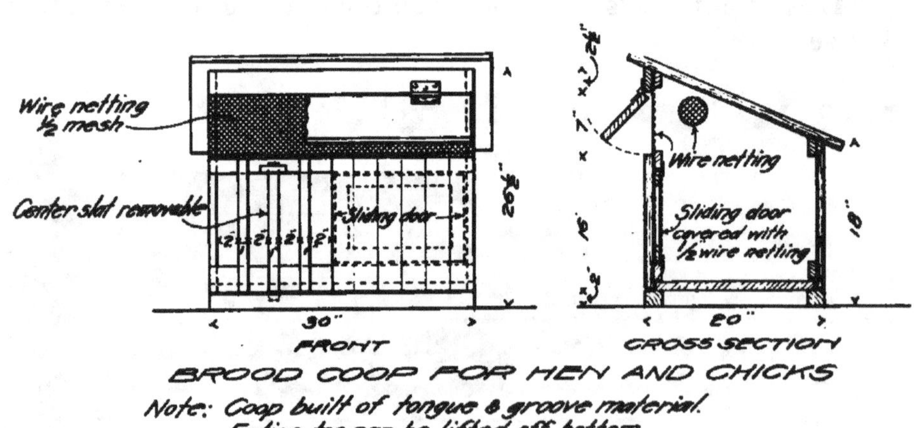

BROOD COOP FOR HEN AND CHICKS

Note: Coop built of tongue & groove material.
Entire top can be lifted off bottom.

Chapter XI

Problems of Incubation and Brooding

WE have reached the most interesting as well as most difficult and important problem pertaining to the poultry industry, viz., the development of the chick. Its development in the egg as an embryo and the aftergrowth of the baby chick are at the very foundation of successful poultry culture. If the foundation is successfully laid there is hope that the superstructure will be enduring. There can be but little profit in poultry culture unless large numbers of chicks can be correctly hatched and successfully reared.

The whole problem goes back to the quality of the egg, and this depends on the character of the foundation stock. This hinges on many factors, such as feeding, breeding, environment, and the personality of the breeder. Many shake their heads and say "There's nothing in it." The trouble is they have not the patience and courage to grapple with its problems. These are not so easy to solve as may appear on the surface. The wisest fail at times.

Producing the Ideal Egg

Take the problem of producing a perfect egg for incubation. Does it mean nothing more than shell, albumen, and yolk? It rather means a shell of ideal shape and texture; egg-contents containing the exact nutrients required to develop the embryo; and an ideal germinal vesicle, one that contains all the possibilities of size, shape, color, vigor, and productiveness required to fulfill the ideal in the breeder's mind.

It is an easy matter to err in the selection of foundation stock. There may be some taint of disease or disqualification that escapes observation, and years of breeding may be necessary to eliminate them. Feeding, environment and heredity determine the character of the embryo. Much that has been said in preceding chapters bears on this problem. When the best possible egg has been produced, if it does not receive proper care its virtue is soon lost.

Conserving the Fertile Egg

1. **Gather the eggs twice daily.** Do not jar or crack them. The vitelline membrane is very delicate and is easily ruptured.

2. **Keep eggs in a room free from drafts,** dampness or bad odors. The room temperature should be 60 degrees, not less than 55 degrees nor more than 65 degrees. At 70 degrees an egg incubates.

3. **Turn the eggs daily.** If they are kept in a 12-dozen case they can be turned by shifting the case to the opposite side. If kept in cabinet drawers with sliding frames, several dozen can be turned with one motion.

4. **Eggs received by express** or parcel post should be allowed to rest for twelve hours to allow the egg structures to be properly composed.

5. **Do not keep eggs** for hatching longer than 10 days if it can be avoided. From this date on the hatchability of eggs vanishes very rapidly.

6. **Do not wash eggs.** After washing they deteriorate rapidly. If they are only slightly soiled the soiled spots can be removed with a damp cloth without wetting the whole egg.

7. **Number each egg** on the large end according to the pen from which it was obtained. In case of trapnesting and pedigree breeding the number of the leg-band should also be marked.

8. It is a good plan to **test all eggs for specific gravity,** using only those for setting that have dense contents.

9. It will pay to **test all eggs with a lamp tester.** This will detect any with thin shells and with cracks, or dents, or containing blood spots, or with yolks adhering to the shell. It will also detect eggs with broken vitelline membrane, in which the yolk becomes fused with the other contents of the egg.

A home made egg-tester will answer all purposes. Using half inch lumber make a box six inches square and twelve inches high. On one side opposite the lamp flame make a hole two inches in diameter. Cover this hole with a piece of leather and in the center make a hole one inch in diameter. If a mirror is placed opposite this hole it will reflect the light through the egg. The top should be covered with tin containing an aperture about three inches in diameter to allow the escape of fumes from the lamp. A picture of such a tester will be found on page 64.

This is the way to handle chicks for profit

Selecting Eggs for Incubation

If pedigree hatching is pursued, eggs from a given hen or a given mating are numbered and only those used with the sitting hen or the pedigree tray, as the case may be. Eggs that should not be set comprise those that are unusually large or small, uneven in shape, round eggs, elongated eggs, stale eggs, dirty eggs, thin-shelled, rough-shelled or mottled eggs. and eggs with low specific gravity. Eggs that may safely be set are those that are ovate in shape with smooth and even texture and of high specific gravity. Specific gravity may be determined by the magic tester, or by weighing with the hand. or by noting the size of the air cell with the lamp tester. If cell is large the specific gravity will be low.

Changes in Incubation

During the process of incubation the air cell gradually increases in size, so that when the chick is ready to be hatched the air cell occupies about one-third of the shell. The yellow yolk is not all used but is retained in the egg to be at the last absorbed into the abdomen of the chick to provide nourishment during the first three or four days of chick life. The albumen is used as nourishment by the developing embryo. The shell becomes soft and easily broken on account of the action of the carbon dioxide.

The most wonderful changes, however, take place in the embryo. When the sperm cell fuses with the germ cell the resulting embryo is but a single, primordial cell, but before the egg leaves the oviduct the embryo has become two cells. known as the blastoderm. At this point development is arrested until favoring conditions occur outside the parent body. Sometimes the egg is retained in the body for a considerable period, incubation sets in, later the embryo dies, and the result is what is known as a stale fresh-laid egg.

The following are some of the changes that take place in the embryo as it progresses toward the fully formed chick:—

At the end of the first day growth is apparent and blood vessels can be seen.

At the end of the fifth day the eye and heart and lungs have made their appearnce with radiating blood vessels.

At the end of the 10th day the bones and muscles are quite well developed.

By the end of the 15th day the skin and feathers are well developed.

On the 18th day the first peep is heard, and on the 19th day the beak penetrates the enveloping membrane and the process of hatching begins.

How the Chick Hatches

Before hatching takes place the embryo must be fully developed. To accomplish this requires a definite number of heat units. When the proper temperature is maintained the embryo receives the correct number of heat units in about 19 days, and hatching then begins. If the temperature is run too high, hatching begins before the 19th day. If run at too low a temperature, the time of hatching is sometimes prolonged to the 24th day. It is a good sign when the hatch comes off on time.

Hatching begins by thrusting the beak through the membrane that encloses the chick into the air space at the large end of the egg. The shell is then pipped. If pipping occurs near the small end of the egg it shows an abnormal condition, and the chick may not be able to hatch. After the shell is

A good hatch

pipped the chick remains quiet for six to ten hours and then begins the work of extricating itself in earnest. The body of the chick is turned in the shell, the membrane enclosing the chick turning with it. As it turns, the shell is pipped until a circular cut is made around the egg. The remaining process is chiefly muscular. By pushing with its feet in one direction and with its shoulders in the opposite direction the two parts of the shell are separated, and the chick emerges into the outer world.

Kind of Incubation

There are two kinds of incubation—natural and artificial. Natural incubation is accomplished by natural means. artificial by human devices. Which is to be preferred? It is not uncommon to hear the following statement: "During the present season I have had poor success. All my chicks were hatched with an incubator. I have lost nearly all of them. Last season I hatched with hens and lost very few."

Two things must be noted, first, that without any doubt a hen hatches a better chick than an incubator and, second, that we must not blame the incubator for all the losses among chicks. If the cause was sifted out it would undoubtedly be found that the losses were due to faulty brooding rather than to imperfect incubation. The art of brooding, notwithstanding all modern improvements, is a long road from perfection.

There is, however, a well-founded prejudice against artificial incubation. Prominent breeders hatch all their breeding stock with hens, believing that the incubator chick has a weakness entailed upon it that in time undermines the vigor of the flock. This prejudice is only increased when a farmer's wife has no trouble in raising 400 to 600 hen-hatched chicks, while her neighbor who hatches many more with an incubator has nothing to show for her labor at the end of the season. There is no reason why an incubator properly constructed and correctly managed should not hatch as good a chick as the hen that dances on her nest and often deserts it at the critical moment. When artificial incubation and brooding are brought to exact sciences this prejudice will disappear.

Natural Incubation

Natural incubation is accomplished by heat from the sun, from fermentation of decaying vegetation, and from the body of the parent. It is usually accomplished by the heat of the hen's body. When a hen becomes broody she enters into a fever (this is denied by some), and her temperature rises from 101 degrees to 107 degrees. This high temperature is communicated to the egg so that during most of the period of natural incubation the upper surface of the egg records a temperature of 106 degrees. For sucessful incubation three conditions are required:

Adequate heat, sufficient moisture and a **supply of oxygen.** All of these conditions are provided by the sitting hen. Her own temperature provides the heat, her body conserves the moisture produced by oxidation within the egg, and the porous nature of the feather allows the access of all the oxygen needed for the growing embryo. There is no excess of heat causing the coagulation of albumen within the egg. There is no excess of moisture causing the chick to drown. There is no lack of oxygen causing suffocation and death in the shell.

Care of the Sitting Hen

The nest is first provided. This may be a box, 12 inches by 14 inches. The depth may be six inches. It is a good plan to put sand or sweet soil in the bottom, and, after properly shaping it, to line the nest with cut straw or fine hay. One or two nest eggs are then put in place, and the hen is secured.

Before placing on the nest she is carefully dusted with insect powder and some of it is also dusted in the nest. It is a good plan to change the hen from the laying house to the nest at night. By morning she will probably be contented with her new surroundings, and the setting of 13 or 14 eggs can be placed under her. Dusting should again be done about the 15th day, and at that time the nesting material should be changed. Occasionally the nest should be examined, and, if any of the eggs have been soiled by a broken egg, they should be washed in tepid water and new nesting material provided. If this is not done the embryos will smother, as the albumen from the broken egg completely closes the pores of all surfaces that it smears.

Where Shall the Hen be Kept?

Not in the laying house, for that means almost certain disaster. There are three good methods of providing room for the sitting hen. One method is to make the nest in the brood coop. This is provided with an outside slatted runway, so that she can secure any needed exercise, and a place is provided where food can be placed before her without interference from other fowls. When the hen hatches, the nesting material is removed, and the coop becomes the home of the hen and brood. It should be stated here that the hen should be kept confined for at least two weeks. After that she may be allowed to range with the brood, at least in the afternoons.

A second method is to provide a small building with compartments. I have used this system for several years. The building is 12x24 feet. There is an aisle through the center and on each side two rows of compartments. The lower floor should be concrete and the upper floor wood. The rooms are three feet wide and four feet deep. This provides for 32 rooms in the building, and, if one hen is placed in each room, that number of hens can be accommodated. Two can be put in each room, if they are taken from the same breeding pen.

In that way 64 hens covering 800 to 900 eggs can be handled with little trouble. If the nest boxes are six inches deep there will be trouble when the hatch comes off, for some of the chicks will creep out of the nest and, unable to get back, will become chilled. This danger is avoided by making a few frames of 1x4 inch boards the exact size of the nest box and setting one on each nest box as the hatch comes due. Only a few of these will be needed as only a few hens are usually set at one time. The building described above is very useful in many ways, providing a place for fattening market fowls and for conditioning birds for the show.

The feed kept before sitting hens consists of equal parts of shelled corn, whole wheat and hulled oats. Water is provided in cups and the feed in small boxes upon the walls. The third method is by a large room with nests around the walls. The hens are confined on the nests and allowed to come off for feeding once a day. A second visit is necessary to see that the hens get back on the nests in good shape. When the chicks hatch, as many as 25 can be given to a single hen. The remaining hens are returned to the laying house as it is not best to set them a second time. Chicks from special matings should be toe-marked or leg-banded so as to preserve their identity. See toe-mark chart on page 162.

Artificial Incubation

For the commercial plant or the breeder who handles the heavy meat breeds or the non-sitting Mediterranean breeds, artificial incubation is almost a necessity. The incubator is his boon.

There are mammoth incubators, accommodating as many as 600,000 eggs, and baby incubators, designed for only 50 eggs. There are hot water machines and hot air machines. Some are heated by coal stoves, others by oil burners, gas, or electricity. As to whether hot water or hot air produces the best chicks, opinion is divided. A hot air machine with adequate provision for moisture seems to have the preference. An incubator, to have any consideration at all, should provide for the three conditions of successful incubation, viz., heat, moisture and ventilation. The incubator that fulfills these conditions in the same degree as the mother hen is the one for which poultry keepers everywhere are searching. For the average farm, the incubator that approaches nearest to the above requirements is the one to purchase.

It is a safe rule to follow explicitly the directions of the manufacturer in setting up, regulating, and operating the machine.

The following rules are of general application and may be helpful:

1. Set the machine in a cellar, where an even temperature can be maintained, or in a room with a firm floor, where the air is pure and the temperature about 60 degrees.

2. Make sure that the machine is level. If not, the high corner will be hotter than the low corner, resulting in an uneven hatch.

3. Do not put the eggs in the machine until it has been regulated so at to run at an even temperature of not less than 102 degrees.

4. If the machine has been used before, give it a thorough cleansing and disinfection. A weak solution of creolin in hot water makes a good disinfectant. Every surface of the interior of the machine should be treated, for the germs of coccidial and white diarrhea often lurk in the incubator and infect the whole brood at hatching time. If there are nurseries lined with burlap, new material should be used, and all trays should be carefully disinfected.

5. Keep the temperature at 103 degrees during the first two weeks. During the third week the temperature should be 104 degrees, except that during the period of hatching no harm will come if the temperature reaches 105 degrees. Remember that the temperature cannot go below 90 degrees or above 107 degrees without seriously injuring the quality of the chicks that may hatch.

6. About the fourteenth day look for a sudden rise in temperature. This is due to the animal heat generated by the developing embryos. It amounts to several degrees, ranging from 4 degrees to 10 degrees. Unless the machine is watched at this time and regulated so as to make allowance for this natural increase, the hatch may be injured. Temperature can be regulated by the size of the flame and by adjusting the thermostat.

7. Begin turning the eggs on the third day and turn twice a day until the 18th day. Turning is accomplished by removing some of the eggs and shuffling gently. Do not turn eggs after the eighteenth day.

8. At the end of the 18th day close the machine and keep it closed until the hatch is completed. Watch the hen. She sits close while the chicks are hatching. When the hatch is completed, remove the tray and put the chicks in the nursery. This will be about the end of the 21st day.

9. Do not cool the eggs until after the seventh day. Watch the hen. She sits close during the first week, hardly leaving the nest for food. Beginning with the eighth day, cool the eggs once a day up to the 18th day. Many chicks are ruined by too much cooling. If we are so careful that the tender, baby chick shall not become chilled, why should we not consider the tender embryo in the shell? Cool the eggs gradually. When they feel slightly cool when applied to the eyelids, then is the time to return the trays to the incubator. Keep the incubator closed while the eggs are cooling.

10. Keep the bulb of the thermometer on a level with the top of the eggs, but it should not touch any egg. It will then register the temperature of the air in the brood chamber. Every thermometer should be tested for accuracy.

11. Keep the room temperature at 60 degrees. You cannot maintain the proper temperature in the incubator if the room temperature goes below 50 degrees or above 70 degrees.

12. Provide an abundance of pure air. A crowded, ill-ventilated living room is not the best place for an incubator.

Why do Chicks Die in the Shell?

Many chicks die in the shell because the germ is weak. They develop until the hatching period and then have not the vital energy to get out of the shell. Many are infected with disease germs to which they succumb before they are able to pip the shell. Some are drowned because of too much moisture. The air cell should be watched to make sure that the eggs are properly dried out at hatching time. Some die for want of moisture. The membrane surrounding the chick becomes dried to the outer membrane, and the chick is unable to turn in the shell. Some die because overheated. The albumen of their blood becomes coagulated by the excessive heat and death ensues. Some become chilled, which means a loss of vitality. Some perish on account of rough handling. The vitelline membrane becomes ruptured or the shell becomes cracked, and in either event death ensues.

What is the Cause of Cripples?

A very common cause is excessive heat in the incubator. It is a very rare thing for a hen to hatch a cripple. She does not permit the temperature to rise above 107 degrees. Cripples arise when the chicks are not able to turn in the shell or when held in one cramped position for a considerable time. Uneven temperature during the period of incubation is the cause usually assigned for these deformities.

Brooding

Brooding, as well as incubation, may be natural or artificial. Natural brooding, with the hen, solves the problem of heat in an inexpensive way. It is a common thing on the farm to set a number of hens at the same time that the incubator is set. When the incubator comes off the chicks are given to the hens, until each hen has about twenty, and then they are put out in coops. It has been found that incubator chicks do not do as well under hens as chicks hatched by the natural method. They do not do as well as under an artificial system of brooding. When large numbers of chicks are to be raised, the incubator and artificial brooder are a necessity. For the average farmer, better results will be obtained by the natural method, for there is no question but that the hen-hatched and the hen-brooded chick is superior.

Artificial Brooding

There are several types of brooders on the market. The fireless brooder depends upon the heat generated by the chicks. When the chicks are a month old or older this brooder can be used to advantage. It can also be used for smaller chicks in a room provided with artificial heat. The feather brooder is one type of this kind that is quite popular. The ordinary box in which the brood chamber is heated by a lamp is used when small numbers of chicks are handled. Some of these give good results. The small colony house provided with universal hover, the heat being furnished by a lamp on the outside of building, is not always satisfactory as the lamp, though enclosed, is exposed to more or less draft. The stove brooder seems to have solved the brooding problem better than any other device. The stove is heated by hard or soft coal or kerosene, according to the type. It is provided with a regulator so that an even temperature can be maintained. The hover throws the heat down upon the backs of the chicks and, as there are different zones of heat from the stove outward, the chicks are able to find the zone of heat adapted to their needs, and there is but little danger of overheating. Some hovers are provided with outside curtains or enclosures, but as a rule they are open. The oil heater seems to have some advantage over the coal heater as it maintains a steady heat and is not so liable to go out.

The Baby Chick

When safely hatched the tender chick is just ready to begin the struggle of life. There are many dangers ahead. To avoid these requires all the resources and skill of the poultry keeper. Nature has provided it with sufficient nourishment

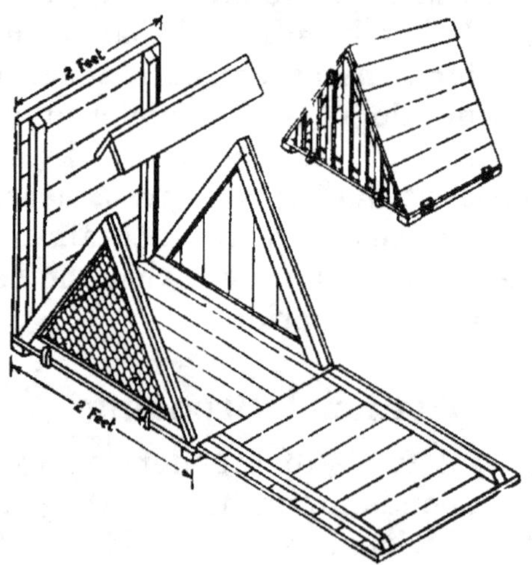

Plan of brood coop for hen and chicks

for 72 hours in the yolk, but recently absorbed into its abdomen. While the chicks are drying and cooling in the nursery of the incubator, the brooder should be gotten ready. If a box brooder, the lamp should be cleaned and filled with fresh oil, the wick trimmed so as to make an even and round flame, free from sharp points, and the hover is cleansed and disinfected.

If a stove brooder, it is thoroughly cleaned so as to remove any rust or soot that would intercept the draft. The pipes are examined, and any in bad condition are replaced with new sections. A hole in the pipe interferes with the draft. The fire is built and the heat regulated so that a thermometer test indicates a temperature of 100 degrees in the outer zone of the hover.

Everything is now ready for the chicks. Only sound chicks are brought to the brooder. Cripples and others that were helped out of the shell should not be saved. This is a time when sentiment must be waived for the common good of the brood. There is this consolation that, if not killed, they would die a lingering death, but the chief reason is because they are probably infected with disease germs and thus a menace to the whole hatch. A merciful way to kill a chick is to pinch it under the wings between the thumb and forefinger. After the chicks have been in the nursery twenty-four hours, they are then transferred to the brooder in a warm basket, every

Another convenient brood coop

precaution being taken to keep them from becoming chilled. They are placed under the hover at once. To prevent any danger of chilling and to get them acquainted with their foster mother it is customary to place a guard around the outside of the hover during the first day. This can be made of muslin or wire screen or even of boards. After the first day or two it can be safely removed. No feed should be given on the first day. On the second day milk can be set before them. At the beginning of the third day they have been out of the shell 72 hours and it will be safe to give them solid food. This should be rolled oats or pin head oats fed in shallow litter. Sweet milk and oats constitute their ration during the first week. On the sixth day after being placed in the brooder, a dry mash is given in a self feeder on an elevated platform. The chicks will soon find their way to it. Use the mash outlined in Ration No. 1 in Chapter 4. The mash, oats, and milk constitute the ration until the end of the second week. At that time a change is made to ration No. 2. The change is made gradually, and at the end of the fourth week change to Ration No. 3. If milk is available throughout the feeding period, the amount of meat scrap in the mash may be reduced 5 per cent. The mash can be fed moist if desired.

After the second week, green or succulent feed should be supplied daily. The milk is given in the forenoon, as it will always be sweet at that time. At noon it is removed, the vessel cleaned, and water given in the afternoon.

A cool room adjoining the brooder room is always recommended as an essential of successful brooding. This room is connected with the brooder room by an opening or a hinged door. If there is but one room, the brooder is placed in one corner or end of the room so that the other portion can be used as a cool room for feeding and exercise. The floors should be covered with two inches of clean, gritty sand, and upon this is placed about an inch of clean, bright, short-cut alfalfa or alsike clover. Chaff from the barn or straw stack is usually unsafe as it contains the spores of Aspergillus. Short-cut rye straw or wheat straw might answer if bright and clean. The scratch feed can be thrown in the litter in the cool room, but should never be fed in litter contaminated with the droppings of chicks or other filth. A good plan is to provide a feed box three feet square and two inches deep. The feed is thrown in litter upon the bottom of this box. When the chicks are through feeding, the box is removed and in that way is kept reasonably clean. This will answer while the chicks are small.

The temperature under the hover can be gradually reduced during the first week from 100 degrees to 95 degrees; in the second week from 95 degrees to 90 degrees; and in the third week to 85 degrees. If the weather is mild the chicks can be weaned from the stove brooder when they are four weeks old and removed to a colony house with fireless brooder. A very simple affair will answer for the fireless brooder, a frame, with legs at corners, covered with flannel or muslin and a border around the sides.

Chicks should never be allowed to become chilled, for a chilled chick is a ruined chick. This happens when they are unable to find their way to the brooder or when the lamp or stove fails for want of fuel or attention.

When chicks become cannibals, or toe-pickers, it indicates a lack of protein in their feed. The dry mash recommended will help cure the habit. It is claimed that direct sunshine contributes to the formation of this habit. If the lights are given a coat of white paint it will help matters.

Key to Successful Brooding

The following is a summary of the essentials of successful artificial brooding:—

1. An even and adequate temperature; a room without drafts and comfortable.

2. An abundance of light but without direct sunshine.

3. Plenty of room for exercise; a shallow litter to encourage exercise.

A popular type of brooder and colony house

4. An opportunity to get out upon mother earth and into fresh air as often and as early as weather conditions permit. In case of long confinement, place green sods upon the brooder floor.

5. A system of feeding that meets the demands of the growing chick; not too much variety at first; sharp sand for grit is better than prepared grit; no feed for the first 72 hours after hatching; no underfeeding, no overfeeding. Green feed and a dry mash after the first week.

6. Extreme cleanliness; this means cleaning the brooder room frequently and replenishing the litter; it means clean vessels, frequently scalded; it means an elevated platform for the mash and water fount.

7. Immediate removal from the brood of any chicks that show symptoms of disease. This is the only road of safety.

8. Fresh air in abundance. Look after the ventilation.

9. A constant warfare upon lice and mites. Avoid fumes that would injure the chicks.

10. A dry floor. Concrete, covered with sand, and then litter is the best.

11. An adjacent room or a portion of the brooder room set apart as a cool place for exercise and feeding.

12. The house must be vermin proof. The runway to the outside should be closed at night to protect against cats, rats and other vermin.

Artificial brooding presents one of the most difficult problems connected with poultry culture. It is the great stumbling-stone of the industry. It is hoped that the few suggestions given above will be helpful.

Chapter XII

From Producer to Consumer

MUCH study and effort are given to production but often the marketing end of the industry is neglected. To buy the best, to produce the best, and to sell well should be the aim in poultry culture.

Disposing of Surplus Stock

There are four possible ways in which the producer can dispose of his surplus poultry:

He can sell direct to the consumer.
He can sell to the huckster who visits the farm.
He can sell to the local dealer in the village.
He can sell to the city commission merchant.

The first method insures a higher price and greater profit for it eliminates one or more middlemen. This method would be used to a greater extent were it not for the difficulty of finding the purchaser and the frequent impossibility of supplying his needs promptly and regularly. Occasionally a hotel or restaurant can be found that will appreciate the opportunity of purchasing direct from the producer. On one occasion I sold a large number of springs to a Chicago restaurant. They were crate fattened, dressed and shipped as required, and the deal proved profitable to both parties.

The second method saves the expense of delivery, but frequently the price is not satisfactory, and unless the huckster is known there is some risk in encouraging him to visit the premises.

The third method, of selling through the local merchant, is common and usually satisfactory, but the local dealer must have his profit, and before the product reaches the consumer it must pass through several hands all of whom must exact a toll.

The fourth method is perhaps one of the best. If an honest commission man is selected, the producer is enabled to get a larger share of the final selling price of the poultry than when he sells to a local dealer.

From what has been said it will be noted that the sale of poultry is not usually a producer-to-consumer transaction. The huckster, the local dealer, the public carrier, the city commission merchant, the inspector, the conditioner, the jobber and the retailer may all have a hand in bringing the finished carcass to the table of the consumer, and each is entitled to pay for his services. The consumer pays and the producer loses. If the farmer properly conditions his fowls and sells to the consumer, either direct or through his commission firm, it would mean larger profit for him and a reduced price to the consumer.

What About the Commission Merchant?

Undoubtedly he performs a valuable service for the producer. If found to be honest, competent and prompt in service he should be cherished as a friend. The dishonest dealer—fortunately there are few of this class—should be forsaken as soon as discovered. Write Prairie Farmer's Protective Union, Chicago, Ill., for names of reliable commissionmen.

Story of a Farmer

An Illinois farmer shipped a coop of choice spring chickens to a commission firm in Chicago. The gross weight at the farm was 194 pounds, the coop weighed 54 pounds, making a net weight for the fowls of 140 pounds. After considerable delay he received returns for 117 pounds at 19 cents per pound, which was the minimum price on the market for that day. The shrinkage in this case was 23 pounds, which, valued at the selling price, meant a loss of $4.37. Under average conditions the shrinkage should have been not more than 10 pounds.

What Happened?

Several things could have happened with this shipment to cause such a loss, viz., failure to water and feed the fowls before cooping; faulty condition of the coop, allowing the escape of one or more fowls; overcrowding, which may have caused the suffocation of one or more; work of "light fingers" among employes of the express company or the commission firm; dishonest weights at the receiving end; slow action on the part of the carrying agent.

The Story of Another Farmer

Another Illinois farmer shipped his surplus poultry to a certain commission firm for a number of years only to discover that the shrinkage on each coop was from twenty to thirty pounds. Satisfied that he was not getting a square deal, he changed to a new comission firm, and a close comparison of weights for a term of years showed that the shrinkage was never more than ten pounds and often there was no shrinkage whatever.

How He Won Out

The farmer made good by observing the following suggestions:

He watered and fed all fowls before cooping.

He made a memorandum of the number of fowls in each coop and their net weight. This was for comparison with returns from the sale.

He made sure that the coop was strong and made secure for the journey.

He avoided overcrowding; not more than twenty average fowls in an average coop.

He provided food for the journey, especially when fowls were to be cooped over night. This was planned so that when the fowls appeared upon the market they could be sold with empty crops. This plan reduces shrinkage to a minimum.

He kept tab on his commission merchant. This is often overlooked. The very few firms that are crooked make it necessary to watch all. Quoting Miller Purvis, an authority on poultry wisdom: "The reputation of the commission merchant should be carefully looked into before making consignments as the cities are full of swindling commission merchants, although there is no lack of honest ones if the trouble be taken to find them."

What to Market

The following classes of market fowls can be disposed of in season. To hold any class longer than the demand continues or market conditions warrant means a waste of feed and a loss of profit.

1. Springs. These are young cockerels and pullets held over from the crop of the preceding season and are put upon the market in the spring months when the prices are high. It will pay to condition them.

2. **Broilers.** These are young cockerels and pullets which are usually marketed in June and July at a weight of two pounds to three pounds. As broilers they may command a better price than if held over till the following winter.

3. **Roasters.** These are the surplus cockerels held over till September. They are put on a fattening ration and sold for fall consumption.

4. **Culls.** These are fowls which are culled from the laying flock, usually in September. If conditioned they command good prices.

5. **Capons.** These are desexed cockerels, which attain large size and are in great demand at fancy prices. They are conditioned for the holiday trade, or in February for the late winter market.

Selling Purebred Stock

The keeper of purebred poultry has sources of income that the mongrel breeder does not enjoy. He can sell his eggs at prices much better than the general market affords. Purebred stock has several outlets, as **day-old chicks, breeding stock,** and **show birds.**

Day-old chick industry. Those who have had experience in selling baby chicks know that the demand is always greater than the supply. This industry has grown to mammoth proportions. There are incubators in use that will accommodate more than 500,000 eggs. Orders are always booked in advance. Shipments are made in specially prepared cartons of corrugated paper. It is customary to give each chick a drink of tepid water before shipping, but no food is given. Fed chicks invariably perish.

The bottom of box is provided with cut straw; the size of the compartment is reduced if the number of chicks is smaller than the capacity; in very warm weather extra openings are made for ventilation if necessary; and chicks are shipped when one day old, not later if a long journey is before them.

Where do the hatcheries get the eggs? It is customary to establish purebred flocks among near-by farmers and these furnish eggs at better than market prices. These flocks are generally free range flocks sustained at a high standard. The purchaser of chicks from the large hatcheries has the advantage of securing stock from farm produced eggs and hatched under scientific, up-to-date methods.

Breeding stock. In every purebred flock there are always surplus cockerels and pullets that can be sold at good prices for breeding purposes. Only the best are saved for this purpose, the inferior stock going to market.

Show birds. These are extra fine specimens, and they are always in demand for exhibition or for special matings in line breeding.

Fabulous prices are sometimes paid for birds of outstanding quality.

Breeding stock and show birds are shipped in light crates. Before shipment they should be gone over carefully to make sure that there are no disqualifications. Feet and shanks should be washed and all dirt removed from under scales. When two or more male birds are shipped to the same address they should be put in separate coops or partitions should be placed between them.

All this requires advertising that producer and consumer may be brought together.

Advertising. Unusual care must be exercised that all advertising may be done with wisdom and discretion. Fortunes have been sunk in advertising. It is folly to use large display ads in poultry journals when the quantity and quality of the stock in possession do not warrant. If the breeder is a beginner or has a limited surplus it would be far better to use a small classified advertisement in the local paper or a good farm paper or in his poultry journal. This will usually sell his surplus and the expense will be but a trifle. If the breeder has a large surplus backed by superior quality he can well afford to launch out into a more expensive advertising campaign, and that means a display ad in the poultry journal, for only through that source can fancy prices be obtained.

Selling Eggs

The value of eggs produced in the United States in 1919 was approximately $750,000,000. The number of chicken eggs produced on the farms was 1,656,267,200 dozens; 35 per cent of these were consumed on the farm. Considering that an egg is very fragile and easily damaged, it must require an enormous expenditure of labor and money to transport this wonderful output from the nests on the farm to the tables of the consumers. The season of large production covers only a few months of spring and early summer.

Were it not for the cold storage plants there would be a glut on the market in the season of high production, and prices would be disastrously low. Storage eggs may have a tendency to lower the prices of fresh eggs in winter, but the fresh egg is always in demand and remunerative prices are maintained in spite of the release of cold storage stocks.

The Chicago Mercantile Exchange gives the following rules for grading market eggs:

1. Eggs shall be classed as **fresh gathered, storage packed,** and **refrigerator.**

2. Eggs shall be graded as **extras, firsts, ordinary firsts,** and **dirties.**

3. The term "**loss**" comprises all eggs that are rotten. broken (leaking), spots, broken yolked, frozen (split), hatched (blood veined), and sour. Very small, very dirty, cracked (not leaking), badly heated, badly shrunken, salted, and chilled eggs shall be counted one-third loss in all grades excepting "Seconds," "Dirties," and "Checks."

This rule applies to the grading of eggs when they are to be sold in the shell, and does not mean that such eggs designated as "loss" (except rotten eggs) shall not be used for canning or drying purposes when same are of a sweet or wholesome nature.

4. **Fresh gathered extras** shall be free from small and dirty eggs, and shall contain fresh, reasonably full, strong in body, sweet eggs as follows:

February 15 to May 15	90%
May 15 to Oct. 31	80%
Oct. 31 to Dec. 31	70%
December 31 to February 15	80%

The balance may be defective in strength or fullness, but must be sweet. There may be a total average loss:

September 1st to June 1st	½ dozen per case
June 1st to September 1st	1 dozen per case

5. **Fresh gathered firsts** shall be reasonably clean, of good average size, and shall contain fresh, reasonably full, strong in body, sweet eggs as follows:

February 15 to May 15	70%
Balance of the year	45%

Mr. Osburn on his way to town with eggs. Eggs for shipment are first packed in cartons, then in the crates. They seldom break when shipped in this way

The balance, other than the loss, may be defective in strength or fullness, but must be sweet. There may be a total average loss:

September 1 to June 1	2 dozen per case
June 1 to September 1	1½ dozen per case

If the loss exceeds this amount by not over 33⅓ per cent, the eggs shall be good delivery upon allowance of the excess.

6. **Fresh gathered ordinary firsts** shall contain the following percentage of fresh, reasonably full, sweet eggs:

February 15 to May 15	60%
Balance of the year	30%

The balance, other than the loss, may be defective in strength or fullness, but must be sweet. There may be a total average loss:

September 1 to June 1	3 dozen per case
June 1 to September 1	1½ dozen per case

If the loss exceeds this amount by not more than 33⅓ per cent, the eggs shall be good delivery upon allowance of the excess.

7. **Dirties** must be of useful quality, sweet flavored, and must not lose over one and one-half dozen per case, loss to consist of rots, spots, and checked or cracked eggs—checks or cracks to count three for one. No. 2 Dirties may be off-flavored, not musty, and must not lose over three dozen per case.

Storage packed eggs are eggs put up for storage, and refrigerator eggs are eggs that have been in storage. The grades of these two classes are not given, as the producer is interested chiefly in "fresh gathered eggs."

Outlet for Eggs

The following are some of the methods of disposing of eggs:

1. **Selling to the huckster.** This saves time and labor in delivery.

The purchaser comes to the farm door and pays cash for his purchases. In remote sections and in busy seasons this is a great advantage, and the farmer can well afford the sacrifice in price.

2. **Selling to the country merchant.** In this event payment is usually made in merchandise. If the merchant breaks even on the eggs he still has opportunity for profit on goods given in exchange.

3. **Selling to private customers.** This method is becoming more popular each year. Sometimes the private customer is secured in the town where the producer does his trading, sometimes in the great city through the help of personal friends. Shipments are usually made by parcel post. A good method of packing is by using the Humpty-dumpty case. Each egg is first wrapped in paper and then packed in suitable cardboard container holding a dozen eggs. After the case is lined with strong paper to prevent leakage, the containers are packed snugly, and after proper labeling, delivered to the post carrier. There should be no loose eggs in the package as these are the ones that are usually broken.

Where eggs are shipped to private customers great pains are taken to select eggs uniform in shape, size, and color, and perfectly fresh. The retainer is returned at the producer's expense.

4. Selling to the commission merchant. This is usually done by the local merchant, but if eggs are produced in large numbers it is advisable to ship in 30 dozen cases to the city merchant.

5. Selling eggs for hatching. A setting of eggs is counted as 15. Eggs for hatching can be shipped by express or by parcel post. The safer method is by express as the package receives more careful handling and goes through with less breakage. If eggs were shipped by parcel post in baskets instead of in cartons, the results would undoubtedly be more favorable.

Packing Eggs for Hatching

The best method is to pack in a basket with strong handle. The basket is first lined with heavy paper, and a thin layer of excelsior is placed in the bottom. Then each egg is wrapped first with paper, then with excelsior, then packed in the basket. When the first layer of eggs is placed they are covered with excelsior and upon this the second layer is placed. When all the eggs are in snug position they are covered with a layer of excelsior, and over this is sewed a muslin cover. A label marked "EGGS FOR HATCHING" should be pasted on the muslin, and a tag containing the shipping address and name of sender is fastened to handle. Eggs packed in this way can be shipped safely either by express or parcel post.

Guarantee. When eggs are sold for hatching a special price is expected. This ranges from $1 per 15 to $1 per egg. It is customary to give a guarantee to protect the purchaser in event of failure to secure a good hatch. What constitutes a good hatch? Not less than eight sound chicks. If the setting cost $1, each chick would cost only 12½ cents, and the same quality of chick would cost about 20 cents from the hatchery. If the setting cost $3 the chick cost would be 20 cents, and the same grade from the special breeder would cost not less than 50 cents. It must be remembered that home hatched chicks are usually superior to incubator chicks that are compelled to undergo the ordeal of a long shipment.

Even though the first cost may be greater, there is less risk in securing new stock through eggs for hatching than in any other way. What should the guarantee be? It is customary to guarantee eight sound chicks. If the hatch falls below this number due to a fault in the eggs, the breeder should replace the infertile eggs free of charge. If the hen forsakes the nest, chilling the eggs, or dances on the nest, breaking eggs, the purchaser will be considerate enough not to ask a rebate under such circumstances. In all business transactions there is more or less risk, and there is no reason why the purchaser should not bear his share as well as the breeder. Where the loss is due, not to the eggs, but to their treatment in the hands of the purchaser, it is only right that any replacement should be entirely optional with the breeder. In the case of high-priced eggs, it should be a rule to replace all infertile eggs without question, if a test is made by the tenth day. Most purchasers are on the square and will not take advantage of any guarantee to secure extra eggs without cost. Rules for selecting eggs for hatching are given in Chapter VI.

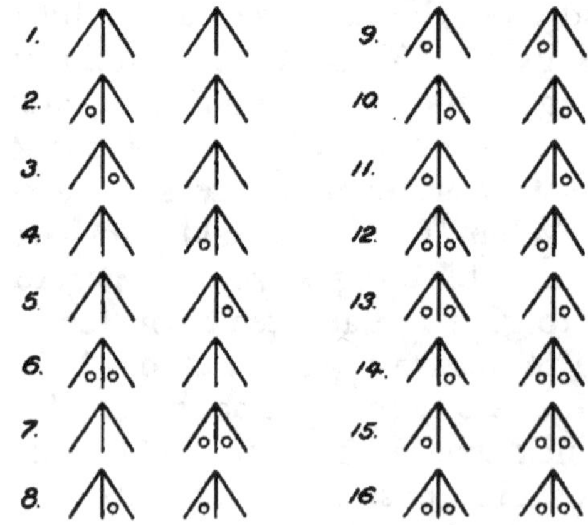

Showing how to toe mark the baby chicks

Chapter XIII

Poultry Sanitation—Pests and Parasites

NINETY per cent of all financial disappointments in poultry raising can be traced to epidemics of disease; therefore a study of the prevention, diagnosis, and cure of disease is important.

What, then, is the road to success? It is not the route of fine buildings and expensive equipment or pedigreed stock, but rather the highway of a healthy flock.

What is health? It is a condition in which the tissues and organs of the organism function in harmony. If there is friction anywhere there is an abnormal condition and disease. Health in a fowl is indicated by a good appetite, sprightly carriage, glossy plumage, a bright full eye, a bright red comb, normal droppings, and pronounced stamina.

What is disease? It is a condition in which the cells, tissues and organs of the organism fail to function in harmony. If only one structure fails to function normally, it affects in a greater or less degree all the other structures, and we conclude that the body is diseased.

How Can Disease be Prevented?

Prevention is better than cure. If disease can be warded off it saves loss of time and money and waste of food that always follow in the train of every sickness. To prevent disease should be the aim and study of every poultry keeper. What are some of the methods of prevention? Here are a few.

1. **Breeding for physical stamina.** The beginner is occasionally wrecked on the shoal of fancy feathers. It is possible to combine fine plumage and vigor, but, if a choice is to be made between the two, physical stamina should have preference.

2. **Prompt isolation.** The spread of disease can be prevented by prompt isolation. An epidemic often can be averted by the quick removal of one sick fowl from the

flock. There should always be a hospital or pest house on every farm where a hundred chickens can be kept. This may be only a room with compartments or suitable coops. It should be located at some distance from the other buildings. If dry, well ventilated and kept clean and comfortable it will be the means of saving many fowls and prove a good financial investment. It is sometimes urged that it does not pay to doctor sick fowls, that they should be killed as soon as discovered. If the same principle were applied to human beings the race would become extinct. The truth is that there is no farm animal that responds more promptly to treatment than the fowl. The poultry attendant must be the judge and, if he finds that a fowl is beyond the help of remedies, the more quickly it is dispatched the better. In the majority of cases, however, good care and correct remedies mean immediate recovery.

3. Removing the causes of disease. The causes of disease have been discussed briefly in preceding chapters. It will be sufficient at this time to enumerate the more important causes:

(a) Overcrowding: Allowing more in the house than one for every four square feet of floor space.

(b) Lack of ventilation: Failing to supply and to distribute an abundance of fresh air.

(c) Drafts: Permitting holes and cracks in the building; failing to provide partitions in long buildings, which are always drafty unless this precaution is taken.

(d) Lack of sunlight: Darkness and gloom are the friends of disease; sunshine and oxygen are the great germ killers.

(e) Dampness: Health and dampness cannot abide together.

(f) Uncleanliness: Disease germs revel in unclean surroundings. Frequent and thorough cleaning and disinfection are necessary.

(g) Lack of exercise: A watch allowed to run down and stand idle will corrode; the chicken is a machine that must have exercise in every part to maintain a healthy condition.

(h) Improper feeding: The feeding of one grain continually or the constant feeding of an unbalanced ration invariably brings disease.

(i) Disease germs: These are brought to the flock in many ways but more especially by domesticated and wild animals. Rats are disease carriers, also English sparrows. They go from farm to farm, spreading contagion. This explains why there are occasional outbreaks of disease even though the flock is kept under ideal conditions.

(j) Weather conditions: Damp and cold windy weather often bring trouble to the flock unless there is adequate protection. This is in evidence in the fall of the year when the young stock are still roosting on the ground or in crowded coops unprotected from sudden weather changes.

What Should be Done When the Fowls Get Sick?

To know the nature of the disease is half the cure. This is often a difficult task. Our knowledge of poultry ailments is incomplete and, even in some well known diseases, the symptoms are so obscure and the lesions so complicated that an error in diagnosis is not improbable. In identifying diseases, two things must be considered: External symptoms and internal conditions.

How to Make a Diagnosis

First, note external symptoms. It is not always possible to determine a disease by external symptoms alone. The same symptom may occur in several diseases. For example, lameness occurs in tuberculosis, liver disease, gout, bumble foot, rheumatism, coccidiosis, and other affections. Diarrhea occurs in cholera, enteritis, coccidiosis, liver disease, vent gleet, and mineral poisoning. It is only by comparing all external symptoms with conditions found in the autopsy that a correct conclusion can be made. Symptoms should be studied carefully. Note whether there is lameness, diarrhea, swellings and where located, peculiarity of action, blindness, coughing, rattling, ruffled plumage, loss of appetite, empty or full crop, or emaciation. In the case of diarrhea, the color and consistency of the excreta should be observed. After a record is made of outward signs, then a dead bird should be examined.

Second, Make a post mortem examination. In making an autopsy, the following tools are needed: A sharp knife, a pair of strong shears, small forceps, small scissors, a dissecting needle and a pair of bone forceps or, instead, a pair of small tin-shears. The dissecting needle can be made by forcing the head of a needle into a wooden handle.

A board, two feet square, should be secured, and this is covered with paper. The specimen is laid upon the board, back down. Sever the skin holding the thighs to the body, spread out the legs and tack them to the board. This will hold the body in position while further work progresses. Cut through the skin and flesh just below the point of keel. Lift the keel and at the same time use the shears and make a cut on each side from point of keel through skin and ribs to the shoulder. By cutting through the caracoids with the tin-shears the breast can be removed. Now make a longitudinal cut from point of keel to vent, cutting through skin and muscle, being careful not to cut the intestines. Spread the skin outward and tack to the board on each side. A full view is now given of the viscera, with no feathers to interfere in the examination. Abnormal conditions may appear at first view, but a critical examination should be made of all the organs. Note the heart. Is it enlarged, or the pericardium filled with liquid? Examine the liver. Are there any spots or discoloration? Is the liver enlarged or is it shrunken? Note the intestines, whether they are inflamed or discolored or enlarged. Examine the kidneys lying in the sacral region. So also, other organs and systems should receive careful study. If worms are suspected it will be advisable to remove the digestive tract and place it upon a clean sheet of paper. It can then be slit from cloaca to proventriculus.

Now, with the facts at hand, both of external symptoms and internal conditions, it will be possible to make a reasonably accurate diagnosis.

After the diagnosis, what? The remedy must be determined and applied. Some assistance along this line can be obtained from Chapter XIV. If a more exhaustive study is desired there are valuable books on poultry diseases, such as those of "Blair," "Dr. Salmon," and "Pearl, Surface, and Curtis" which can be secured from any poultry journal company. The following list contains remedies that will be useful for all kinds of livestock and many of them are valuable for family use.

Poultry Remedies

Epsom Salts, $MgSO_4$.—There is no more valuable drug for the poultry farm than Epsom Salts. It is indicated in liver diseases, rheumatism, gout and many intestinal affections. One-half to one teaspoonful to a fowl, either in mash or water or one pound to 100 hens in mash or water are suitable doses.

Copperas, Iron Sulphate or $FeSO_4$, is a blood tonic and a good germicide, hence its use is indicated in contagious and infectious diseases. One teaspoonful to 100 hens in drinking water is advised.

Castor Oil.—This is useful for affections of the alimentary tract. It can be given with medicine dropper, or in warm milk by drenching.

Bi-carbonate of Soda, Na_2CO_3.—It is useful in cases of fermentation in crop or intestines. Dose: One-half teaspoonful in a quart of water.

Quinine.—A good remedy for colds and cases of chicken pox and roup. It is a good tonic. Dose: A one-grain capsule to an adult fowl.

Calomel is recommended for liver diseases and constipation. Dose: One-fourth grain to the adult fowl.

Nux Vomica is a remedy for leg weakness and indigestion. Five drops in a pint of water is the recommended dose.

Turpentine.—This remedy is good for colds, rattling in the throat, intestinal worms, and it acts on the kidneys. It should be given with sweet oil, one part turpentine to five of sweet oil, one teaspoonful of mixture constituting the dose for a fowl.

Carbolated Vaseline is useful as an ointment in cases of colds, roup, chicken pox, sore eyes, scaly leg and frozen combs.

Permanganate of Potash, $KMnO_4$, is a valuable disinfectant and germicide. Its use is indicated in any contagious disease. A saturated solution should be made in a large bottle. Crystals should be added until the water has dissolved all that it will hold, so that crystals still remain in the bottom of the bottle. From this stock solution the daily supply of water for the flock can be prepared by adding enough of the solution to turn the water a deep red.

Sulfur is useful in preparing ointment, as a remedy for worms, and as a germicide. Sulfur ointment is prepared by adding one teaspoonful of sulfur to one-half teacupful of vaseline or lard.

Strychnine in one-thirtieth grain doses is a stimulant and tonic and is indicated in cases of asthenia and debility.

Mustard.—This is very useful as a tonic. It stimulates and tones up the digestive and reproductive organs and promotes the health of the flock. The dose is one heaping teaspoonful for six hens, or about one tablespoonful for 10 hens. It is usually fed in a moist mash.

Iodine.—Tincture of iodine is used to prevent and destroy infection. For internal use it is recommended in cases of aspergillosis and tuberculosis. Dose, two drops to adult fowl.

Iodoform.—A powerful germicide, used as a powder on abraded surfaces and in preparing an ointment, which is recommended in cases of diphtheria, vent gleet, favus and chicken pox.

Unguentine.—This ointment allays pain and is very healing. It is recommended for burns, abrasions and all exposed surfaces. It can be obtained of any druggist in collapsible tubes.

Disinfectants and Germicides

A disinfectant is any substance that destroys any infectious or contagious matter. The difference between infectious and contagious is only one of degree. A disease is said to be infectious when it is communicated through food or drink or by direct contact with the infectious material. Tuberculosis, bacterial enteritis, aspergillosis, coccidiosis, black head, and vent gleet are infectious diseases.

A contagious disease is of a more virulent nature and is communicated not only through food and water but through the air. Cholera, roup, chicken pox, diphtheria, and favus are contagious diseases.

A germicide is any agent that destroys disease germs, such as bacteria, spores of fungi and animal parasites. Formaldehyde, carbolic acid and creolin are germicides.

Valuable Disinfectants

Copperas.—A strong solution of copperas is one of the best of disinfectants. It removes foul and musty odors from buildings and grounds and is harmless to the flock. It should be sprinkled on the floor every time the building is cleaned.

Slaked Lime.—This should be used often to sprinkle on the drop boards and in damp and musty places. It is one of the best substances to use in disinfecting the ground. It should be scattered on the ground when any contagious disease is present, and when gape worms attack the young chicks it should be scattered freely over their runs and, after the ground is plowed or spaded, another coat should be applied.

Lime Wash.—This is made by slaking a peck of lime with a little boiling water, just enough to keep it covered. When the process of slaking is completed, add enough water to make a thin paste. Strain this through a fine sieve to remove any lumps or foreign particles and then add a solution of four quarts of salt in hot water. When ready to use the mixture add hot water to bring it to the proper consistency. The disinfecting power is increased by adding one pint of carbolic acid to five gallons of wash. It may be applied with a brush or with a spray pump.

Crude Carbolic Acid.—A good disinfecting solution is recommended as follows:—Use one pint of crude carbolic acid, one pound of laundry soap, and one gallon of kerosene. The soap is first dissolved in a gallon of boiling water and then the kerosene and carbolic acid are added. When ready to spray, add water equal to the mixture.

Creolin is a good germicide. Add a sufficient quantity, about two per cent, to the amount of hot water required, and then use as a spray or wash. It can be used to wash out incubators and brooders. As it is more powerful than carbolic acid the quantity required will be very small.

Formalin.—This is a 40 per cent solution of formaldehyde gas. A five per cent solution of formalin is used to spray the interior of buildings, incubators, coops, etc. After the spray is applied, the room or incubator is closed as nearly air tight as possible. When incubators are disinfected they should be allowed to air and dry out before putting in the eggs. This is to avoid any possible injury to the chick germs by the disinfectant.

A good stock dip makes an effective disinfectant. Zenoleum is one of the best.

Parasites and Enemies

Enemies and parasites of the farm flock are found in the soil, in the air, in the water and in the food. If there is a creature that has more guerrillas on its trail than a chicken I have not been advised.

Bacteria are too small to be seen by the naked eye but accumulate in such vast numbers that they are a constant menace to health. There are upwards of 100,000 bacteria in a cubic centimeter of virgin soil; ordinary milk contains 20,000 to a cubic centimeter; and it is estimated that in a milligram of fecal matter from a fowl there are upwards of 30,000,000 of these micro-organisms. This gives us some idea of the vast numbers of bacteria that form a part of the invisible world of living organisms. Many of these are pathogenic and only await an opportunity to prey upon the body. Fortunately for the chicken there are many "anti-bodies," or agents that prey upon disease germs, so that a fowl in vigorous health is able to escape infection. Such are the hydrochloric acid of the gastric juice which acts as a germicide; the serum of the blood which is also germicidal; the white blood corpuscles which attack and destroy disease germs, the liver cells and other cells of the body which engage in constant warfare to protect the body. This explains why healthy fowls are often able to resist an epidemic of disease, while the weaker ones succumb.

Lice.—The common poultry louse, **Menopon pallidum,** attacks all parts of the fowl's body. It can be seen moving rapidly among the feathers. Unlike the mite it lives by biting rather than by sucking. It remains and breeds upon the fowl's body, and often clusters of eggs can be seen in the region of the vent. It causes loss of vigor and emaciation. Probably the best treatment is the sodium fluorid powder recommended by the government. A pinch of the powder, such as can be seized between the thumb and forefinger, is applied to the feathers next the skin on the head, neck, back, under wings and at base of tail. The amount should be limited to about 10 small pinches, as it is irritating if used too freely. It should not be used with sitting hens. One pound will treat a flock of 100 hens and effectually destroy all the lice. Other methods of destroying lice are to dust insect powder among the feathers, using an ointment, such as blue ointment, and by dipping.

If chicks are infested with head lice a good application is sweet oil, which is effective and harmless unless used in too great a quantity.

Red Mite, Dermannyssus gallinæ.—The color of the common mite is gray but becomes red after feeding upon the blood of a fowl. They attack their victim at night. In the daytime they are secreted in cracks and other hiding places where they breed. Knowing their habits it is not a difficult matter to get rid of them. Crude petroleum or processed oil applied to the roosts, nests, and in all cracks and crevices will do the work. A good spray is effective, but should be repeated.

Mites are serious pests. On account of their small size they are often overlooked until great damage has been done. If roosts are supported as recommended in Chapter X the pests will not be likely to make their attacks, because unable to reach the roosts.

Depluming Mite, Sarcoptes lævis gallinæ.—This attacks the skin at the base of the feathers producing an intense itching, which causes the fowls to pull out their feathers. Apply an ointment to the parts affected consisting of one part sulfur and four parts vaseline.

A bad case of scaly leg

Scaly-leg Mite, Sarcoptes mutans.—This parasite works under the scales of the toes and shanks, filling the spaces with a powdery substance which causes the scales to be raised, giving the roughened appearance so characteristic of the disease. The mites can be killed and the disease cured by dipping feet and shanks in crude petroleum, or processed oil, or a mixture of four parts of raw linseed oil and one part of kerosene.

Other pests such as bedbugs, chiggers, ticks, and fleas are at times very annoying and in some sections their infestations are very serious.

Air-sac Mite, Cytodites nudus.—This mite infests the air sacs, occasionally in such numbers as to produce emaciation and anemia. There appear to be no effective means of ridding a flock of this pest. The free use of sulfur is advised. Any substance inhaled into the air sacs sufficiently strong to destroy the mite would probably injure the fowl. If the whole flock is affected it would be best to dispose of them and put the new stock on fresh ground.

Worms

Many species of round worms and tape worms infest the alimentary tract producing a variety of serious diseases, such as anemia, emaciation, epilepsy, enteritis, toxic poisoning and indigestion. Worms lay the foundation for a long train of diseases by destroying the resisting power of the fowl so that

Wry neck

it becomes an easy prey to bacteria and other disease germs. Santonin and male fern are usually recommended as remedies for worms, but they are expensive. The remedies recommended in Chapter XIV will be found effective and far less expensive. They are tobacco, salts-sulfur-copperas, and gasoline methods.

Major Enemies

The following list is by no means complete. Only the more common ones are named.

Cats.—It is possible to train a cat so that it will not harm small chicks. It should be fed regularly. When the first brood comes off place the cat near and watch. If there is any effort to destroy one of the chicks she should be caught and punished severely. She will probably give no further trouble. Cats are sly and cannot be trusted unless well fed and carefully trained.

Dogs.—They must be trained like cats. They are often very destructive to flocks of turkeys. Only good dogs should be kept.

Hogs.—There is nothing more aggravating than a chicken-eating hog. Its example is soon followed and the whole herd becomes like a pack of wolves. Lock them up in chicken-proof pens and remove the flock as far away as possible.

Minks are very cunning. If they can find a hiding place on the premises they will remain for weeks, each night destroying one or more fowls. The fowl is caught behind the head, the blood is sucked, and then the body is dragged away and the flesh consumed, or at least a portion of it. When a mink appears a search should be made. Better use a shot gun than to allow it to escape. If once discovered and frightened it will probably leave the premises.

Weasels are destructive of young chicks, destroying a score or more in a single night and carrying their bodies to some place of concealment. Coops should be made vermin-proof and always closed at night.

Skunks destroy eggs and sometimes attack chicks or fowls. Their nests should be sought and raided.

Rats.—Rats are undoubtedly the most destructive of all poultry pests. They consume and contaminate the feed, they destroy eggs and young chicks, they carry disease from farm to farm and from flock to flock, and they damage buildings and equipment. A pest that destroys property value to the amount of $200,000,000 annually and requires the constant labor of 300,000 farmers to supply it with food should have some attention from our lawmakers. We have a good law which provides for the eradication of the Canada thistle, and legislation encouraging a warfare upon the symptoms of human tuberculosis, and have summoned the nations to discuss the limitation of armaments that war may hide its deformed head for at least ten years, and yet here is a pest that is more destructive than any noxious weed, that is responsible for the spread of tuberculosis to a large extent, and that has caused the loss of more lives than all the wars of history, but it is allowed to go unscathed, tolerated by governments and ignored by legislators.

An impatient and long suffering electorate will some day insist that legal measures shall be adopted to protect the masses of the people from this scourge.

There is some risk in using poison to destroy rats. If put out in the evening all the remnants should be gathered up early the following morning and destroyed to avoid danger of poisoning domestic animals.

Traps are helpful but large numbers should be used. Woven wire traps are sometimes successful. A correspondent from southern Illinois reports that he captured 300 rats during a single summer by this method. The trap was baited and left open until the rats became accustomed to conditions. Then it was closed and a vessel of milk placed inside. Rats are fond of milk and in a dry season can be easily enticed into the trap. A board or sack placed over the trap helps matters.

A well trained ferret will drive out the rats from a farm. A good rat dog will be a valuable help. A campaign of rat extermination on the farm will mean the saving of many dollars. Buildings should be made rat-proof. Remove the harbors and the rats will disappear.

Hawks destroy large numbers of chickens and turkeys annually. The Goshawk, Prairie Falcon, Marsh Hawk, Sharp-shinned Hawk and the Sparrow Hawk are some of the more common varieties. The shot gun seems to be the best remedy.

Crows destroy eggs and young chickens and turkeys. They are difficult to capture, but if one can be shot and hung up in a prominent place near the poultry buildings there will probably be no further depredations from this source.

English Sparrows are pests without one redeeming quality. They consume and contaminate poultry feed. They are the carriers of disease. They destroy the eggs and young of native birds. On account of their rapid multiplication they are becoming a menace to the farm. Sparrow traps are advised for their destruction.

Chapter XIV

Diseases and Remedies

We have come to one of the most vital subjects pertaining to the poultry industry. The six important factors in the maintenance of a healthy flock are isolation, sanitation, disinfection, cremation, diagnosis, and application of suitable remedies. The first three of these items have been discussed in Chapter XIII. The remaining items are considered in this chapter.

Cremation

All dead animals on the farm should be cremated as soon as possible. If allowed to remain where other animals can gain access to them it will result in spreading of disease. Dead fowls can be thrown into the stove, or burned upon a fire in the open made of cobs or old wood, or they can be consumed in a crematory. A common method of making a crematory is as follows: Make a solid foundation of concrete. Upon this there is built a fire box having an inside measurement of one and one-half feet wide by two feet and sixteen inches deep. Across the top of the fire box iron grating, or old wagon tires or iron rods, one inch in diameter, should be laid about one inch apart. Above this is built an oven of brick laid in cement. This should be about 14 inches deep. A cover of galvanized iron is made for the top. This should have a handle in the middle and a flue opening at one end. In front of the fire-box should be an opening to admit fuel and remove ashes.

Burying dead animals is not always satisfactory. If that practice is followed dead fowls should be buried in places removed from the buildings and about three feet deep.

Diagnosis and Treatment

In the following discussion of diseases an effort has been made to give very briefly the cause, symptoms and treatment of the more common affections. It is hoped that they are sufficiently full to aid the reader in determining the character of some of the diseases which affect the flock as well as the appropriate remedy to use.

Causes, Diagnosis and Treatment of Poultry Diseases

I. Affections caused by accidents.

Bumble Foot is caused by a bruise on ball of foot or puncture and infection.

Symptoms: Lameness, swelling of foot, infection.

Treatment: Lance the swelling, wash out with one per cent of creolin or paint with iodine, anoint with carbolated vaseline, cover with sterilized cotton and bind with surgeon's linen or with adhesive tape. Keep fowl in clean place.

Broken bones are the result of accident. If a clean fracture, set the bone, wrap with a layer of cotton, apply thin splints, bind in place with adhesive tape. A cure will be effected in three weeks.

Tears, or rents, are also accidental. Pluck feathers from around the sore, wash with one per cent creolin solution, then sew up rent with silk thread that has been saturated in alcohol. Apply unguentine ointment to all surface sores, burns, etc.

Abscesses occur from bruises or infection. Lance and wash out with a one per cent of creolin, then apply tincture of iodine to edges of sore or sprinkle with iodoform or apply unguentine ointment.

II. Internal accidents

Apoplexy is caused by a rupture of a blood vessel in brain; is sometimes due to excessive fat, or to fright, or to strain in the act of laying. It occurs more frequently with hens than with males.

Symptoms: Fowl falls from roost or dies on nest. Death is sudden without premonitory symptoms.

Treatment: Compel exercise, feed less, give Epsom salts occasionally to whole flock.

Broken Egg. Occasionally an egg becomes broken in cloaca or oviduct. This may be due to treading of male or striking an object in flying from perch.

Symptoms: Straining and bloody discharges.

Treatment: If in cloaca, remove with finger and swab with unguentine ointment or olive oil.

Egg Bound is retention of the egg; may be due to excessive fat or constipation or to excessive size of egg.

Symptoms: Straining, hen goes frequently to nest without laying.

Treatment: Apply hot bandages, inject olive oil into cloaca, use gentle pressure in dislodging the egg.

Prolapsus is caused by the straining used to expel a large egg; sometimes causes an eversion of the oviduct. This is called prolapsus.

Symptoms: Straining and protrusion of the oviduct.

Treatment: Apply hot bandages and olive oil.

Crop Bound is caused by an obstruction in the outlet of the crop or by a paralysis of the muscular walls of the crop due to impaction.

Symptoms: Distended crop, difficulty in swallowing, drowsiness, pale comb.

Treatment: Give a tablespoonful of sweet oil, massage the contents and force as much as possible out through the mouth. In removing the contents by an operation, the incision should be made on upper side of crop. Sew crop membrane and skin separately in closing up the opening.

Obstruction of the pharynx is caused by growths or masses of food which can be removed if discovered in time.

Obstruction of the cloaca is generally due to constipation or an egg-bound condition. An injection of olive oil will give relief.

Internal ruptures, as of the heart, blood vessels, oviduct or kidneys, may be due to an excess of blood with high pressure, to a very large egg in oviduct, or to an accumulation of urates in the kidneys.

Symptoms: Weakness, drowsiness, death.

Treatment: Avoid overfeeding. Feed a balanced ration.

II. Affections arising from abnormalities

1.—Abnormal development

Wry tail. Cause: Probably inherited, though it may be due occasionally to the cramped position of the tail in coops or roosting places. Counted as a disqualification.

MALES WITH DEFECTIVE TAIL CARRIAGE.

1. SQUIRREL 2. WRY

Symptom: Tail twisted or carried to one side.
Treatment: If a natural deformity there is no cure. Birds should be kept in roomy quarters with roosts removed from the wall.

Crooked Back. Cause: Probably inherited in most cases, though some cases may arise from crowding in coops.
Symptom: The bird appears as a hunchback.
Treatment: No cure and such specimens should not be used for breeding.

Crooked Keel. Cause: Improper feeding of chicks. malnutrition, inherited.
Symptoms: Keel bent, twisted or curved. The rocker keel is not a deformity, but indicates large capacity, therefore good fecundity.
Treatment: No cure, a disqualification. Chicks should be supplied with bone-forming material.

Deformed Beak is probably due to some accident or unfavorable environment preventing the proper development of this structure.
Symptoms: The mandibles may be crossed or one mandible is shorter than the other.
Treatment: There is no cure for crossed mandibles, but when they are of uneven length, the longer one can be trimmed to some extent so as to make possible the seizing of food.

Sidesprigs occur on the sides of single combs and are creatures of heredity.
Symptoms: They may occur on the sides of blade or upon the points.
Treatment: They disqualify and should never be removed.

Hypertrophy is the undue enlargement of any organ, such as the liver, spleen, kidney or heart, and may be due to any one of several causes. The more common causes are overfeeding, lack of exercise, malnutrition and bacterial infection.

Dropsy is an accumulation of liquid, serum or water, in heart or abdomen. Dropsy of abdomen is caused by unsanitary conditions and improper feeding; dropsy of the pericardium is associated with other diseases and is probably caused by them.
Symptoms: Sluggishness, rapid heart beat. Dropsy of abdomen can be detected by its swollen condition and it is soft to the touch.
Treatment: Puncture the skin to remove the liquid and then give a good tonic, nux vomica, strychnine, or tonic No. 1.

Atrophy is wasting of the tissues of any organ, due to pressure from excessive fat, or the result of inflammation. Atrophy of the ovary and testes occur as the result of age. Such birds should be conditioned for market.
Treatment: Epsom salts may be helpful in early stages. A strong stimulant, such as mustard, would be helpful.

2.—Abnormal growths

A tumor is a new growth of tissue believed to be caused by the development of dormant or unused embryonic cells. They begin to develop under a given stimulus, which may be bacterial in character or the result of some injury. They seem to affect organs of the body which become inactive by reason of age, so that they seldom occur in fowls under one year of age. The Maine Experiment Station reports about nine per cent of cases of tumor in all autopsies made. The ovary is more commonly affected than any other organ.

Symptoms: Tumors may be benign or malignant. A benign tumor is usually enclosed in a capsule and is harmless because it does not penetrate surrounding tissues. A benign fatty tumor is an illustration. Malignant tumors penetrate the tissues and if removed reappear. They sometimes spread from one organ to another. These tumors attack a number of the tissues of the body.

Symptoms: The fowl becomes sluggish, appetite is poor, and there is emaciation in some cases.

Treatment: The real cause and cure are unknown.

Cancers. Tumors of the epithelial and mucous membrane type are known as cancers. A cancer which involves the squamous epithelial cells of the skin is known as epithelioma. Other types attack the proventriculus, gizzard, intestines, liver, spleen and ovary.

Treatment: Kill and cremate the diseased fowl for there is no known cure.

Internal Abscesses are in most cases probably due to infection. If the cause is removed there may be some cases of recovery.

Corns usually result from bruises and may be caused by narrow perches.

Normal ovary on left. Ovary from hen infected with bacterium pullanum on right

Symptoms: Lameness, ball of foot calloused.

Treatment: As far as possible, without bleeding, remove the corn with sharp knife. Apply tincture of iodine. Make the roosts broad and flat.

III. Affections due to climatic conditions

Sunstroke, or heat prostration, is not uncommon during the extreme heat of summer. It may be due to insufficient shade, lack of water or to hot, ill-ventilated buildings.

Symptom: The fowl falls as if paralyzed.

Treatment: Apply cold water to head, which may be beneficial in mild cases.

Frosted comb and feet. Fowls that roost in the open in rigorous climates are often affected with frozen combs or feet. Lack of ventilation in roosting quarters and access to free range in below-zero weather are common causes. Fowls should be confined in severely cold weather.

Symptoms: Parts are stiff and swollen.

Treatment: Hold affected parts in cold water until the frost is removed. Apply carbolated vaseline to which have been added a few drops of turpentine. Application should be made several times.

IV. Affections caused by improper sanitation and exposure

Common colds or catarrh arise from overcrowding at night, and subsequent exposure to drafts. A draft on the side of the head affects the eye and results in inflammation and subsequent infection.

Symptoms: An offensive roupy odor, swelling of eye, discharge from nostrils, matted feathers under wing.

Treatment: Provide ventilation without drafts and transfer young chicks from coops to permanent roosting quarters, early in the fall. Use a spray and force permanganate of potash into the slit in the roof of the mouth and give permanganate of potash in drinking water. Remove sick birds to comfortable quarters and give each a one-grain capsule of quinine.

Pip is the hardening of the mucous membrane of mouth and tip of tongue, caused by inflammation or mouth breathing when nostrils are closed by colds.

Symptoms: Difficulty in breathing and eating.

Treatment: Open the nostrils and apply glycerine or carbolated vaseline to the hard growth in the mouth.

Bronchitis. Causes: A drafty building, irritating gases, dusty litter.

Symptoms: Difficult breathing. Mucus forms, and young chicks are often strangled; drowsiness; drooping wings; ruffled feathers.

Treatment: Isolate sick birds. Doctor Salmon recommends 10 drops of turpentine in a teaspoonful of castor oil. Repeat the dose for several days. One-fourth of this dose is sufficient for chicks. Avoid dusty litter and irritating vapors.

Congestion of Lungs. Cause: Chilling the surface of the body.

Symptoms: Distension of blood vessels in lungs and closing of air spaces; drowsiness; rapid breathing; dark red or bluish black comb.

Treatment: Disease generally fatal. There is no remedy. Put birds in comfortable quarters and give a good tonic.

Congestion of Brain—Vertigo. Causes: Fright, excitement, blow on head, intestinal worms, indigestion.

Symptoms: There is a rush of blood to brain, fowl staggers, moves in a circle, walks backward and turns the head backward.

Treatment: Give teaspoonful of Epsom salts in water. Remove the cause.

Pericarditis. Cause: Exposure to cold and dampness.

Symptoms: Inflammation accompanied with dropsy of the heart sac; weakness and difficult breathing.

Treatment: Remove causes and give four grains of carbonate of soda. Endocarditis is an inflammation of the inner membrane of the heart and should have similar treatment.

Rheumatism. Cause: Dampness and cold drafts.

Symptoms: Lameness, swellings and inflammation of joints.

Treatment: Remove affected birds to dry room with board floor. Correct conditions in poultry house.

V. Affections caused by poisons

Ptomaine Poisoning. Cause: Eating decayed animal or vegetable food.

Symptoms: Lack of control of the muscles. Comb becomes black, occasional diarrhea; prostration and limber neck; head turning toward breast.

Treatment: A teaspoonful of castor oil in warm milk; or a level teaspoonful of Epsom salts in water. Follow with tonic found in formula No. 1.

Limberneck. Cause: Eating the maggots of the green bottle fly, **Lucilius Caesar**; probably also caused by eating decayed matter and the flesh of fowls that have died of the disease. It is also associated with intestinal worms.

Symptoms: This is not a contagious disease, but is considered a symptom of other diseases. Head hung down between feet, and there are convulsions in which the neck is twisted in different positions. When the head is turned backward and twisted and lies upon the back the affection is called wry-neck.

Treatment: Blair recommends giving equal parts of oil of turpentine and sweet oil, one teaspoonful at dose. Follow this in a half hour with all the sweet milk the fowl can drink. Pure lard, a tablespoonful melted and poured down the throat, will sometimes effect a cure.

Salt Poisoning. Salt is a valuable food. Used in excess it is a fatal poison. Convulsions, prostration, and diarrhea are symptoms. Milk is indicated as a remedy.

Arsenic Poisoning. Fowls contract arsenic poisoning accidentally from rat poisons and arsenical sprays. The symptoms are choking, excessive secretion of saliva, difficult breathing, unsteady walk, convulsions. Milk, white of egg, flaxseed, or sulphate of iron are recommended as antidotes.

Copper poisoning results from careless disposal of spray mixtures containing copper sulphate. The symptoms are diarrhea of blue or green color, prostration, convulsions, paralysis. Milk and white of egg are antidotes.

Lead and zinc poisoning resembles copper poisoning in symptoms. Sulphate of soda is recommended as an antidote.

Ergot poisoning occurs in feeding rye as a sole ration. The ergot of rye is a serious poison and the symptoms are trembling, prostration, and gangrene of comb and tongue.

Quinine in one-grain capsules should be given daily. I have known large flocks to be lost by this affection. The ergot is produced by a fungus which infests the rye. This grain should not be fed to chickens.

VI. Affections caused by improper feeds and feeding and malnutrition

Malnutrition is a failure in digestion and assimilation, so that the fowl does not flourish. It may be due to weakness and failure of the organs of digestion and assimilation, so that they do not function properly, or it may be due to dietary deficiency. Weakness, emaciation and loss of appetite are the symptoms. A balanced ration should be provided and a tonic to stimulate the digestive organs. Use formula No. 1.

Asthenia is known as "going light" and is the result of dietary deficiency or malnutrition, and is often produced by a specific germ. It occurs in flocks where young stock is kept with older fowls in crowded quarters. Those that fail to get adequate food become emaciated. The symptoms are voracious appetite, increasing emaciation, inflammation of the intestines, and constipation. Affected birds should be isolated and given food rich in nutrients and a good tonic.

Sore eyes arise from several causes. They may be a symptom of one of several diseases, such as colds, roup, chicken pox, diphtheria and favus. There seems to be an infectious disease of the eyes not associated with other diseases. The eyelids become glued together and there is an accumulation of pus in the eye. Another cause of eye trouble is the irritation caused by the use of insecticides or liquid lice killers. Still another cause is dietary deficiency. The food lacks in vitamines, or growth principles, and sore eyes is a symptom. Sometimes the eye breaks down and sloughs away.

All affected chicks or fowls should be isolated promptly and kept in dry comfortable quarters. Open the eyelids and remove any accumulation. Wash the eyes with a weak, tepid solution of creolin and apply unguentine ointment which can be secured of any druggist in collapsible tubes. Use permanganate of potash in drinking water and feed a balanced ration with an abundance of green feed

Beri-beri manifests itself in paralysis of the legs, and can be produced by feeding polished rice, or any diet deficient in Water Soluble B. Another name of the disease is polyneuritis.

Treatment: Feed a balanced ration, and give the tonic recommended in formula No. 2. Give an abundance of green feed.

Crop Inflation. Sometimes the crop becomes inflated with gas, the result of bacterial fermentation. A disinfectant, such as permanganate of potash in drinking water, or a few drops of creolin in a quart of water will give relief.

Gout is a kidney disease resulting from failure to eliminate the urates. There are two varieties, visceral and articular.

Visceral Gout. Cause: An excess of one kind of food, especially a food rich in protein, such as tankage or corn. Dampness favors its development.

Symptoms: Lameness. Visceral organs covered with white, chalky deposit, emaciation, but good appetite.

Treatment: Give more variety in food and Epsom salts in mash or drinking water. One pound to 100 fowls.

Articular Gout. Cause: Dampness. Lack of balanced ration. A diet too rich in proteids.

Symptoms: Lameness, swelling of joints in toes, diarrhea in advanced stages.

Treatment: Remove birds to dry comfortable quarters with board floor. Give teaspoonful of Epsom salts every third day.

Indigestion is caused by lack of balanced ration, unsanitary conditions, lack of green food.

Symptoms: Dullness, loss of appetite, diarrhea.

Treatment: Give Epsom salts, one pound to 100 fowls. Follow with a good tonic.

Constipation. Cause: It may follow enteritis, or may be due to the character of the ration.

Symptoms: A dry condition of intestinal tract, and hard dry fecal matter, obstructing free evacuation, dullness, straining.

Treatment: Give level teaspoonful of Epsom salts, remove obstruction in cloaca, give injection of sweet oil and glycerine. Moist mashes and green feed are recommended.

Diarrhea. Cause: It sometimes accompanies indigestion, and is due to overfeeding.

Symptoms: Impacted crop, dullness, excrement whitish, yellowish or greenish, and often watery.

Treatment: One tablespoonful of castor oil to each affected fowl. A small crystal of sulphate of iron in drinking water.

Gastritis, or Inflammation of Proventriculus. Cause: This disease is determined by the kind, quality and quantity of the food.

Symptoms: Inflammation of the proventriculus, dullness, loss of appetite, roughness of feathers and constipation.

Treatment: Give Epsom salts in drinking water, a cooked mash and plenty of green food, and milk to drink will be helpful.

Hepatitis, or Inflammation and Hypertrophy of Liver. Cause: Unbalanced ration, excess of protein in diet, feeding one kind of food.

Symptoms: Inflammation of the liver, loss of appetite, sluggishness, yellowish color to the skin, liver enlarged, tender, and engorged with blood.

Treatment: Epsom salts in drinking water, one pound to 100 fowls, together with an equal quantity of bicarbonate of soda. Give a balanced ration and compel exercise.

Jaundice. Cause: Overfeeding, lack of exercise, decayed or tainted food.

Symptoms: Excessive formation of bile which is absorbed into the blood, giving yellowish color to comb and wattles.

Treatment: One teaspoonful of Epsom salts, good sanitation and balanced ration.

Fatty Degeneration of the Liver. Cause: Ration too rich, resulting in a deposit of fat in the liver tissue.

Symptoms: Similar to hypertrophy of the liver and it is believed to be a stage of that disease.

Treatment: Encourage exercise, give green food and more variety in the ration.

Leg Weakness. Cause: Improper feeding, growth of flesh out of proportion to the bone development, lack of exercise. Board and cement floors are unfavorable.

Symptoms: Leg weakness is the loss of control of the muscles of the legs. There is weakness and the chick sits down to eat; this is followed by loss of appetite.

Treatment: Give less carbohydrate and more protein in the feed, such as meat scrap, granulated bone and milk.

Soft Shelled Egg. Cause: It may be due to a lack of lime or the fowl may be too fat or there may be inflammation of the oviduct.

Treatment: Give plenty of charcoal and grit and green feed. A balanced ration and exercise will cure the disease.

VII. Affections produced by parasites

1. **External animal parasites** cause such affections as emaciation, anemia, feather-pulling, scaly leg. See Chapter XIII.

2. **Internal animal parasites.**

Emaciation. There are three species of round worms that seriously affect poultry.

Ascaris inflexa is about two inches long and is generally found in the fore part of the small intestine. It is yellowish white in color.

Heterakis perspicillum attains a length of two and one-half inches. It is yellowish white and is found in all parts of the intestinal tract. Sometimes it causes constipation by obstructing the course of the food.

Heterakis papillosa is a very small white worm from one-third to three-fourths of an inch long and may be found in any part of the digestive tract, more especially in the ceca, where it accumulates in great numbers. It is believed to be instrumental in inciting the disease among turkeys known as black-head. All these worms produce emaciation and if unchecked result in death. Blindness, limberneck, epilepsy, and emaciation are some of the symptoms.

Treatment: Use one pound of Epsom salts, one pound of sulphur and one-half pound of pulverized copperas, mix carefully, and feed one pound of this mixture in 10 quarts of mash. This should be sufficient for 100 to 150 fowls. Repeat this for five days, once a day, and after that once a week for three weeks. Keep the feeding place clean and well disinfected to prevent the further spread of the infestation.

Epilepsy.—Tape Worms. The symptoms of epilepsy are trembling, jerky movements and convulsions. It is produced by a species of tape worm, known as **Taenia infundibuliformis.** This is about five inches long and is found in the intestinal tract, usually near the ceca. It burrows into the intestinal wall with its head and the segments absorb nourishment from the food in the intestines. The terminal segments are filled with eggs. When these break away they are deposited on the ground and picked up by the fowls and thus the infestation is spread.

Treatment: The New Jersey Experiment Station recommends the tobacco treatment for worms as follows: "For each 100 birds use one pound of finely chopped tobacco stems. Steep in water for two hours and use the tea for the moist mash. The regular dry mash will answer for making the moist mash, feeding one-half the usual amount. Feed lightly during preceding day and nothing at all on the morning of the day for treatment. At two o'clock feed the tobacco mash spreading it out so that each fowl can get a portion. At four o'clock feed a second moist mash of the same quantity as first only add 12 ounces to one pound of Epsom salts. Dissolve in water instead of the tobacco solution. Repeat the treatment in a week."

Another treatment for worms is to use one pint of gasoline in the mash for 100 birds.

Worms cause enormous losses. They are responsible for a long list of diseases and must be fought persistently.

Blackhead. Cause: It is produced by an animal parasite known as **Amoeba meleagridis.**

Symptoms: Circular yellow spots on liver, enlarged ceca plugged with cheesy matter, and enlarged liver. Drowsiness, loss of appetite, drooping wings and tail, head turning to a dark color, constant diarrhea.

Treatment: Isolate all infected fowls. Mix sulphur, Epsom salts, sulphate of iron and quinine, equal parts of each, in two-grain capsules. Two capsules constitute a dose and should be given twice daily. Keep sulphate of iron in drinking water and sprinkle sulphate of iron crystals in any stagnant pools about yards. Disinfect roosting places, remove droppings and sprinkle ground with lime. The success in checking this disease will depend on the promptness with which diseased birds are isolated and dead birds cremated. Similar treatment should be given in case of an epidemic among chickens, one capsule to dose.

Coccidiosis. Cause: It is produced by an animal parasite known as **Coccidium tenellum.** It attacks turkeys, geese and chickens.

Symptoms: Yellow diarrhea, ceca plugged with yellowish pasty matter, loss of appetite, drowsiness, head becomes a scarlet red in early stages. In geese the kidneys are affected, and the fowls become prostrate, turning upon the back. Emaciation and death result.

Treatment: Isolate affected fowls and disinfect premises. Use two per cent solution of creolin, or a strong solution of copperas. Scatter lime about the roosting places. Give a teaspoonful of castor oil to chickens, and double the dose for turkeys and geese. Follow the oil with a capsule of Formula No. 1.

Coccidial Diarrhea attacks baby chicks and resembles white diarrhea. A whole brood becomes infected, probably in the incubator. Infected broods should be isolated from other chicks. The sick chicks should be removed from the brood and destroyed as observed. The brooder room should be thoroughly disinfected with a solution of copperas and fresh litter supplied frequently. A teaspoonful of castor oil in a quart of warm milk should be given to drink in the forenoon and about two grains of copperas in a quart of water in the afternoon.

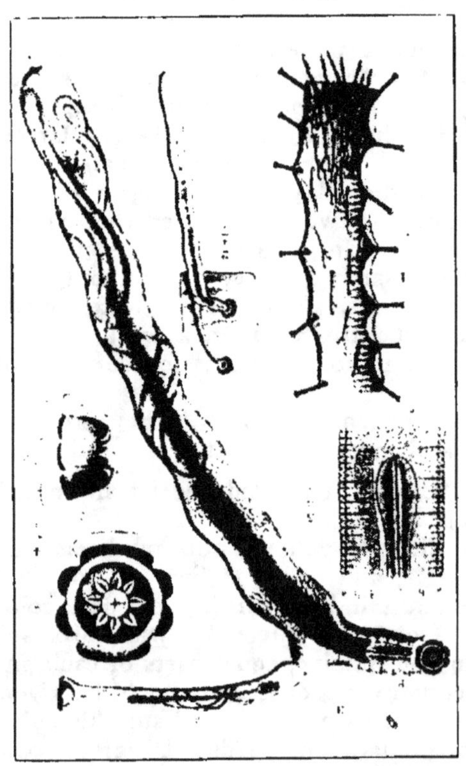

Showing full size gape worm. Also section of trachea with worms attached

Air-Sac Mite, Cytodites nudus, crawls through the nostrils and attacks the bronchi and air sacs of the body. When present in large numbers it produces stupor, emaciation, suffocation and death. The infestation will go through a whole flock unless checked. Give sulphur in the food and keep the premises disinfected.

Gapes. The gape worm, **Syngamus trachealis,** is especially fatal to young chicks. The eggs are obtained from infected soil, from earth worms and by eating the worms coughed up by infested chicks. Turkeys harbor these parasites and communicate them to chicks.

Symptoms: Gaping, sneezing, coughing, and discharge of mucus.

Treatment: Remove the chicks to uninfected ground. Disinfect all vessels. Cremate all dead chicks. Use permanganate of potash in drinking water. Scatter lime over infected ground. Spade or plow and then give another dressing of lime. Take two horse hairs, tie at ends and cut off projecting portions close to knot and insert into the trachea through the larynx. Give the hairs a twist and withdraw, removing the worms by the operation.

3. External Vegetable Parasites

Chicken Pox. The specific germ of chicken pox has not been isolated. Some consider it an animal parasite and some a vegetable parasite. Some think it is a form of roup. Others consider it a blood disease. Small wart-like nodules of a greasy appearance appear on comb and face. Eyes become affected and are sometimes destroyed. A high fever, thirst and stupor develop. It is estimated that 50 per cent of all cases die.

A case of chicken pox

Treatment: Disinfect buildings and premises with creolin solution. Put sick fowls in comfortable quarters and add a few drops of creolin to the drinking water. Give one grain capsule of quinine to each fowl, and after removing the scabs from head, apply iodoform-vaseline ointment.

White Comb or Favus is caused by a fungus known as **Achorion Schonleinii.** It attacks the comb or face and occurs first as white powdery scurf or as white patches, sometimes round and sometimes irregular in shape. These spread until the whole surface of comb, face and wattles is covered and then the disease invades the feathered regions of the neck. At first there is no noticeable effect upon health, but as the disease progresses the fowl weakens, loses flesh and may die. A fowl with favus should never be allowed in the breeding pen. Infected buildings should be disinfected to destroy the spores and supplied with fresh litter.

Treatment: Apply an ointment of iodoform and vaseline, using one-half teacup of vaseline and as much iodoform as can be placed upon a dime; or apply an ointment of one part of red oxide of mercury to eight parts of vaseline. Give a good tonic or a one-grain capsule of quinine. Favene is a good proprietary remedy.

4. Affections Caused by Internal Vegetable Parasites

Aspergillosis is caused by several species of mold, **Aspergillus,** the more common being **Aspergillus fumigatus.** The spores of this mold occur on straw and grain.

Symptoms: Fever, rapid respiration, rattling in throat, diarrhea, emaciation, spots on liver and patches in mouth.

Treatment: Disinfect frequently and supply clean litter, free from dust and musty odor. Isolate all affected fowls. It is claimed there is no cure. Probably the best treatment is tincture of iodine, allowing two drops to each fowl. It can be given in a tablespoonful of water. Follow up for several days.

Brooder Pneumonia. Aspergillosis attacks the young chicks causing enormous losses. Watch the litter. Keep everything scrupulously clean, disinfect frequently, isolate diseased chicks promptly, and use the iodine treatment, five drops in a pint of drinking water.

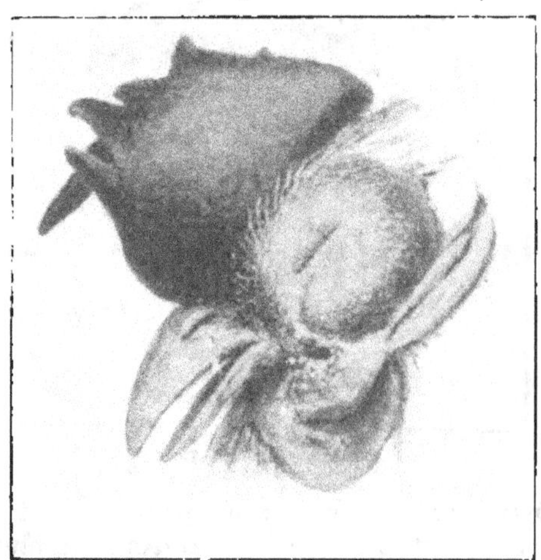

A case of roup

Roup is a disease of the respiratory organs. The specific germ producing it has not been discovered. It is very contagious. The eyes. nostrils, larynx and trachea are affected.

Symptoms: Swollen eyes, discharge from nostrils, foul odor, stupor, emaciation, difficult breathing. Roup entails weakness upon the offspring for several years. Therefore, no fowl that has had roup should be used as a breeder.

Treatment: Isolate all infected fowls promptly. Treat as follows: Give permanganate of potash in drinking water. Dip the head of each sick fowl in a strong solution of permanganate of potash and on the following day apply an ointment of iodoform and vaseline to all parts of the head, and give a small quantity internally. An ointment of lard and sulphur can be used instead. As a tonic give a one-grain capsule of quinine, daily. If there is an accumulation under the eyelids remove it carefully. A good method of applying permanganate of potash is to force it into the slit in roof of mouth by means of a hand spray.

Diphtheria is considered an advanced stage of roup. It is sometimes called canker. Best authorities believe it is a distinct disease, though the earlier symptoms resemble those of roup. It is very contagious.

Symptoms: Stupor, sleepiness, difficult breathing, head swollen, false membrane in mouth and throat, which becomes a thickened yellowish, cheesy mass, as the disease advances.

Treatment: Where the disease is far advanced it is well to kill the fowl and cremate the body. Isolate promptly, and treat with iodoform ointment as advised for roup. Give permanganate of potash in drinking water and supply soft food. Give tonic, Formula No. 1, or one-grain capsule of quinine.

Influenza resembles a severe cold or a mild epidemic of roup. It is probably caused by a specific germ.

Symptoms: The same as a cold and accompanied with diarrhea.

Treatment: Isolate sick birds, give a dose of Epsom salts and follow this with a one-grain capsule of quinine.

Thrush is a fungus disease attacking the œsophagus, the specific fungus being **Saccharomyces albicans.**

Symptoms: Violent convulsions. White patches in the œsophagus, and sometimes in mouth.

Treatment: Wash the mouth with a 10 per cent solution of borate of soda and give a good tonic.

Pneumonia generally follows congestion of the lungs. The serum of the blood escapes through the capillary walls, and coagulates in the air spaces. The disease is undoubtedly caused by a specific germ.

Symptoms: Ruffled plumage, dark comb, rapid respiration, loss of appetite.

Treatment: Avoid exposure by removing sick birds to a warm room. Use 10 drops of aconite and bryonia in each pint of drinking water. Give soft nourishing food, a little at a time.

Tuberculosis. Cause: This destructive disease is caused by a specific germ, **Bacillus tuberculosis.** The organs usually affected are the liver, spleen, intestines, mesentery, and occasionally the lungs. It is transmitted by means of infected birds and animals. It is believed that the specific germ which produces this disease in man is the same as that of birds and mammals, the different types, human, bovine, and avian tuberculosis, being due to environmental adaptation. Nevertheless these types are transferable. A fowl eats the sputum of a human tubercular patient, contracts the disease and dies. A rat eats the fowl, and also contracts the disease and dies. A hog eats the rat, and becomes diseased. Further the rat contaminates food of human beings and farm animals and in that way communicates disease. A human being drinks the milk of a tubercular cow and in that way becomes infected. Thus the vicious circle is continued. It is an endless chain. English sparrows, without doubt, are instrumental in communicating this disease. It is evident that it will be a hopeless task to eradicate tuberculosis unless radical measures are taken to destroy the cause. We need clinics on the farm. There should be state-wide campaigns against vermin which harbor and spread the plague.

Symptoms: Lameness, pale comb, emaciation, bright eye, good appetite, tubercles or raised nodules on liver, spleen, intestines, or mesentery. A post mortem examination always should be made.

Treatment: Destroy the entire flock, or if the disease appears to be restricted to a few birds, kill all that are emaciated, and isolate all the others, giving each fowl two drops of tincture of iodine daily in drinking water. Tincture of iodine is used effectively in incipient cases of human tuberculosis and should give favorable results with fowls. Renovate and disinfect the buildings and premises, and keep a close watch for new cases.

Infectious Leukaemia is a blood disease produced by a specific germ known as **Bacterium sanguinarium.** There is noted a decrease in the red corpuscles and an increase in the white corpuscles.

Symptoms: Increased temperature, drowsiness; the heart, intestines and blood are pale. An increased number of leucocytes.

Treatment: Improved sanitation and a tonic as recommended in formula No. 2. (Page 27.)

Sleeping disease is an affection of the blood caused by a germ known as **Streptococcus capsulatus gallinarum.**

Symptoms: Sleepiness, lameness, swollen eyes, pale comb.

Treatment: Improved sanitary conditions, and give a good tonic.

Vent Gleet, or Cloacitis, is an inflammation of the cloaca, probably produced by a specific germ. It is infectious, being transferred from one fowl to another in copulation.

Symptoms: Inflammation of cloaca, white diarrhea, inflamed skin around the vent, foul odor.

Treatment: Cut away the feathers around the vent and wash with a one per cent solution of creolin. Then make a swab by wrapping cotton around the end of a stick, cover this with iodoform ointment and swab out the cloaca. One treatment will probably effect a cure. Keep the sick birds in dry comfortable quarters.

Bacillary White Diarrhea is a disease of chicks produced by a specific germ, **Bacterium pullorum.** A hen that has had white diarrhea in its early life will transmit the disease through the egg to the chick. The germs of the disease become localized in the ovary, and when the yolks develop the bacteria become incorporated in the egg and are thus communicated to the embryo during incubation. Only a few of the chicks that hatch may have this disease, but it is quickly communicated to the remainder of the brood. The contagion spreads through droppings.

Symptoms: Stupor, rough feathers, emaciation, loss of appetite, whitish discharge, pasting up behind, sharp cries from chicks, and they act as if chilled, and keep close to the hover.

Treatment: Begin with foundation stock and cull out all weaklings. Dip eggs in alcohol, thoroughly scrub and disinfect the incubator, using four per cent solution of creolin. When chicks are hatching darken the incubator so they cannot pick at the droppings. Remove from the brooder any chicks that show signs of the disease. This must be done very promptly and the chicks should be destroyed and their bodies burned. Keep the feeding floor scrupulously clean and scald all vessels. Add to the milk and water for all chicks of an infected brood about five drops of creolin and one teaspoonful of Epsom salts to each quart of fluid. Good sanitation may prevent the spread of the disease, but there is little hope of cure when it has reached an advanced stage.

Bacterial Enteritis or Mixed Infection is an inflammation of the small intestine and is often of bacterial origin. It may be caused by toxic poisons produced by worms, or from eating paint skins, salty meat, or other poisons. If it is produced by infection, due to filthy conditions and tainted food, the cause must be removed. It is then known as bacterial enteritis.

Symptoms: Diarrhea, loss of appetite, comb pale to dark.

Treatment: Disinfect. Give each sick bird a teaspoonful of Epsom salts and follow with a good tonic. Give one pound of Epsom salts to 100 fowls in drinking water. After which give permanganate of potash in drinking water as long as the disease persists.

.This disease is a true mixed infection, as several species of bacteria are known to contribute to its existence in the flock.

Dysentery is a bacterial inflammation of the large intestine.

Symptoms: Diarrhea, the discharges often being bloody. There is loss of appetite and drowsiness.

Treatment: The same as for enteritis.

Cholera is a contagious disease produced by a minute bacterium, **Bacillus avisepticus.** The disease spreads rapidly and is fatal in most cases. The bacteria are found in the blood of infected fowls.

Symptoms: The urates which are normally white assume a yellowish tinge; afterwards the droppings become a bright yellow and in the final stages turn to a bright green. There is drowsiness, ruffled feathers, loss of appetite, thirst, fever and pale face and comb. An autopsy shows inflammation of the digestive organs, kidneys, and mesentery.

Treatment: Renovate and disinfect buildings and yards frequently. Remove and promptly destroy all infected birds. As a general treatment for the flock give one teaspoonful of creolin in three gallons of drinking water. Epsom salts should be given once a week until the disease disappears. If sick birds are to be treated, give creolin in drinking water as advised and tonic No. 1 (page 26). Remember that this is a highly contagious and incurable disease and all infected birds should be far removed from the healthy flock.

Chapter XV

The Poultry Account

DOES poultry pay? Very few are able to answer in the affirmative, or at least answer the question "How much does it pay?" because no record is kept and no balance struck at the end of the year. A poultry account book will help answer the question. The method here proposed requires an inventory at the beginning of the year and at the beginning of each quarterly period thereafter, viz., January 1, April 1, July 1, and October 1. A summary is also required at the end of each quarter, also an annual summary at the end of the year. The latter will show just what the poultry business has done during the year, and the quarterly summary will show the profit or loss during the preceding quarter.

Schedules

The schedules outlined in this chapter can be copied into a book of convenient size, and the system can be used as successfully as if a book with printed forms were available. They are based on the requirements of the income tax schedule and cover the following topics: Inventories, Income, Expenses, Quarterly Summary, Annual Summary, Daily Egg Record, Incubation Record, Losses, and Orders.

Inventory

At the beginning of each quarter a count should be made of all stock and products on hand. The number and quantity of each should be entered in the proper column as shown in Schedule No. 1. In determining the amount to be used in the "value" column it is advised to use the average market value of stock in the preceding year. If an inflated value is used, and losses occur during the quarter or year, then the summary will make a false showing. Under present, 1921, conditions the following prices seem fair and safe; $1.50 each for cocks and cockerels; $1 for hens and pullets; $5 for toms; $3 for turkey hens; $1 for ducks; $4 for geese; and 50 cents each for guinea fowl. In the case of eggs the average market value for the quarter is advised. Whatever prices are adopted for stock, these should be retained through the whole year. If a change is made with every variation in the market more or less confusion will result. The quarterly inventory will take care of all losses during the quarter at the average price, for, when the inventory is taken, it will not include any dead fowls, and, therefore, the "value" column will show a deduction accordingly. Buildings and equipment may or may not be considered in the quarterly inventory. They are taken care of in the annual inventory, and their repair and depreciation count as deductions in estimating the final status of the business. The inventory for the first day of any year or month will be the same as for the last day of preceding year or month. Why make a quarterly inventory? Because it enables the poultryman to keep in close touch with his business and brings him face to face with losses and leaks. It stimulates an interest in the poultry end of the farm enterprise and furnishes an exhibit of the condition of the farm flock. Why should the poultryman wait till the end of the year to discover whether he is playing a winning or a losing game? The quarterly summary tells the tale. If losses have occurred he will discover them and can plan to avoid them in the future.

Income

The American hen should be given credit for all she produces. It is not fair to make all sorts of demands upon the products of the industry without giving due credit for every item of production. This means that due credit should be given for all sales of stock and products, all eggs and stock used for household or given away, all eggs used for incubation, and all feathers and fertilizer sold or used in garden or field. If credit is given for eggs used for incubation they should be estimated at market price, and it is advised to place only a nominal value (not full value) on the under-month chicks at hand when the inventory is made. The price of the egg has gone into the chick. If the value of the egg is five cents and there is a 50 per cent hatch, the egg-cost of each chick is 10 cents. But its real value is not less than 20 cents. Probably, therefore, 10 cents is a fair estimate of the value of a baby chick for the inventory in case credit is given for the eggs used in incubation. In one month, however, this value has increased to 20 cents, in two months to 35 cents, and in three months to 50 cents. At three months it is well past the danger point and has reached a marketable age. If you give the hen credit for all she produces, then you can know definitely whether the industry is worth while.

Expenses

The first cost of buildings and permanent equipment should not be entered in the expense account. It would hardly be fair to charge against the income of any one year the cost price of buildings, fences, incubators, brooders, and other major equipment having a long life-period. This is charged off in the annual inventory as depreciation, the percentage of depreciation being determined by the life-period of the equipment. Minor equipment, such as water founts, feeders, crates etc., which usually last only two or three years, should be charged off as an expense at the time of purchase and therefore no depreciation should be entered against them. Purchased labor should be entered as an expense, but the operator's labor may or may not be counted as an item of expense. If not, his wage will be represented in the net income at the close of the year. If desired, however, he can keep a labor record of the number of hours actually expended on the poultry enterprise, and, charging a reasonable rate per hour, he can enter this labor as an expense in the annual summary.

Keeping Feed Costs

In determining the amount of home-grown feed consumed in a given period it is a good plan to weigh or measure the daily ration of such feeds for several days and from these data make an estimate of the average daily amount consumed. Knowing this, the amount consumed for the month or quarter is easily determined. This estimate will be sufficiently accurate for all practical purposes. To be strictly accurate it will be necessary to weigh or measure the home-grown feeds every day, or they can be weighed in bulk and kept separate from the feeds of other livestock on the farm. In the Expense Schedule, No. III, there is a line for totals. These totals are determined by adding together all the expense items. When the totals are all brought to this line it is simply a matter of addition to determine the total expense for the quarter. The same rule applies to the schedule for incomes, No. II.

Quarterly Summary

When the gross income and total expenses for the quarter have been determined and the inventory for the end of the quarter has been taken, then, by using Schedule No. IV, it is a simple matter to determine the quarterly gain or the quarterly loss as the case may be.

Annual Summary

The status of the year's business is determined by using Schedule No. V. The total of all quarterly net losses deducted from the total of all quarterly net gains (incomes) should give the net income for the year. But there may be other receipts than those included in the quarterly schedule, viz., from insurance, sales of equipment, show premiums, etc., and these should be added to the total of quarterly net incomes.

So, also, there may be added to the total of quarterly net losses all repairs and depreciation of buildings and other items of expense not included in the quarterly schedules, such as interest on borrowed money, taxes, insurance, etc. A separate list of these expenses should be made and entered elsewhere in the record under Schedule XII.

An annual summary based on the requirement of the income tax schedule should be made as outlined in Schedule No. V. The data for this summary can be obtained from the statements of income and expenses as found in Schedules No. II, No. III, No. IV, No. X, No. XI and No. XII.

Daily Egg Record

Schedule No. VI is used to make a record of all eggs gathered and disposed of each day. The eggs may be gathered from separate pens or from the general flock, or both. They may be given away, devoted to table use, sold on the market, used for incubation, or used to fill orders. This record will be of great value if faithfully and accurately made.

Incubation Record

A record of eggs set is of value for reference. From this record it is possible to determine the date of hatching, the value of the eggs used in incubation, the fertility of the eggs, and the age of a chick when it has reached maturity. Refer to Schedule No. VII.

Losses

Schedule No. VIII is used for recording all losses that occur. It is advisable to keep such a list, even though these losses are taken care of in the inventory. It will be a silent reminder of the disappearance of profits.

Orders

A record of all orders received will save many a heartache. The record should be as complete as possible and should include the following data: Date received, name and address of purchaser, shipping station, items ordered, conditions, price, date paid, and date shipped. Refer to Schedule No. IX.

Depreciation

The percentage of depreciation is determined by the life-period of the building or equipment. A building with stone or concrete foundation and shingle roof will have a life period of 40 years. The depreciation on the first cost should be 2½ per cent annually. If the building sits upon the ground and has a paper roof its life-period is reduced to 20 years. The depreciation should be 5 per cent. All major equipment should receive a depreciation of 10 per cent. In all cases, depreciation should be estimated on the original cost, else the building will become perpetual and the cost will never be charged off. Schedule No. XIII shows the method of recording depreciation.

Sundry Schedules

Schedule No. X is for recording all purchases made of stock and eggs.

Schedule No. XI shows a record of all miscellaneous income not included in Schedule II.

Schedule No. XII gives a record of all miscellaneous expenses not included in Schedule III.

These items are repairs, insurance, interest, taxes, show expenses, etc.

A few pages set apart for memoranda will not come amiss

Value of the System

The use of the system recommended in this chapter will be found practical and comprehensive. It is a matter of great satisfaction to know just what any farm enterprise has accomplished. The annual summary tells the tale, but this cannot be made unless the entries are made faithfully throughout the year. On the average farm one minute each day or ten minutes at the end of each week will be sufficient to make all entries. Complete the records at the end of each quarter and make out the quarterly summary, then at the end of the year it will be a small task to prepare the annual summary. Try it. You will be pleased with the knowledge acquired and probably surprised that the outcome is so favorable.

A Few Things to Remember

1. Failure to keep an account is to conduct the business on a guess.
2. A system of accounting shows up the strong points and the weak points of the poultry enterprise.
3. To give the poultry a square deal requires that credit be given for every item of production. Thirty-five per cent of poultry products are consumed on the farm and the hen should have due credit.
4. It is the waste that eats the profits. Every pound of feed represents an outlay and should be turned into finished products, and every egg should be conserved.
5. The poultry account reveals losses, discovers profits, warns of danger, points out the safe course, and stimulates to highest endeavor.

Schedule No. I—Inventory

YEAR	On Hand Jan. 1		On Hand April 1		On Hand July 1		On Hand October 1	
Items	No.	Value	No.	Value	No.	Value	No.	Value
Buildings:—								
Laying and Breeding Houses...								
Incubator House............								
Brooder House...............								
Colony Houses...............								
Conditioning House..........								
Supply House................								
Other Buildings.............								
Major Equipment:—								
Fencing.....................								
Incubators..................								
Brooders....................								
Grain Sprouters.............								
Cabinets....................								
Heating and Lighting Systems .								
Office Supplies.............								
Other Items.................								
Feeds and Supplies:—								
Grain.......................								
Mill Feeds..................								
Concentrates................								
Green Feeds.................								
Oyster Shell................								
Grit and Charcoal...........								
Other Supplies..............								
STOCK								
Chickens:—								
Cocks.......................								
Hens........................								
Cockerels...................								
Pullets.....................								
Chicks......................								
Eggs........................								
Turkeys:—								
Toms........................								
Hens........................								
Poults......................								
Eggs........................								
Ducks:—								
Drakes......................								
Ducks.......................								
Ducklings...................								
Eggs........................								
Geese:—								
Ganders.....................								
Geese.......................								
Goslings....................								
Eggs........................								
Other Fowls:—								
Feathers and Fertilizer:—								
Totals......................								

Schedule No. II—Income

DATE	FOWLS: All varieties, Adult and Young				EGGS			
	Number and kind sold	Amount	Number and kind consumed and given	Value	Number and kind sold	Amount	Number and kind consumed, incubated, given	Value
Totals:								

Schedule No. III—Expenses

DATE	FEEDS				SUPPLIES		MINOR EQUIPMENT		LABOR	
	Amount and kind purchased	Cost	Amount and kind home-grown feeds used	Value	Amount and kind purchased	Cost	No. and kind	Cost	Time	Cost
Totals:										

Schedule No. IV—Quarterly Summary

QUARTER ENDING ____________

Gross income for quarter..............................	$	
Inventory value at end of quarter.....................		
A = Total income plus Inventory........................... =		$______
Total expense for quarter............................	$	
Inventory value at beginning of quarter...............		
B = Expenses plus Inventory................................		$
A–B = Net gains for quarter...........................		$
B–A = Net loss for quarter...........................		

Schedule No. V—Annual Summary

For the Year Ending ________________

1.—Inventory of stock and products, End of Year.............	$	
2.—Sales of stock and products during year—Schedule II.......		
3.—Stock and products given away and used—Schedule No. II..		
4.—Miscellaneous receipts—see Schedule No. XI		
5.—Total: Items 1, 2, 3 and 4..............................		$
6.—Inventory of stock and products, beginning of year.........		
7.—Cost of stock and eggs purchased during year—Schedule X..		
8.—Total: Items 6 and 7..................................		$
9.—Gross profits for year; Item 5, less item 8 =...............		$
EXPENSES		
10.—Expenses (Schedule III), not including item 7..............	$	
11.—Repairs of fences, buildings and major equipment—Schedule XII......................................		
12.—Depreciation of fences, buildings and equipment — Schedule XIII......................................		
13.—Other expenses not included above........................		
14.—Total expenses: Items 10, 11, 12 and 13..................		$
Net Income or Profit for Year: Item 9, less item 14....		$
Loss for Year: Item 14, less item 9 =..................		

Schedule No. VI—Monthly Egg Record

Month of ________________, 19___

Eggs Gathered							Eggs Disposed of					
Day	Pen 1	Pen 2	Pen 3	Pen 4	Pen 5	Flock	Eggs sold		Eggs used		Eggs set	
							No.	Price	No.	Value	No.	Value
1 2 3 to 31												
Totals:												

Schedule No. VII—Incubation Record

QUARTER ENDING ________________ 19___.

Date	No.	Hen or incubator	Location	Number and variety of eggs	Value	Fertility	Hatch
Totals:							

Schedule No. VIII—Losses

YEAR ____________

DATE	ADULT STOCK		YOUNG STOCK		OTHER LOSSES	
	No. and kind	Value	No. and kind	Value	Items	Value
Totals:						

Miscellaneous Schedules

Schedule No. IX—Orders

QUARTER ENDING ________________ 19___.

ORDERS RECEIVED AND FILLED

Date.—No. 1

Rec'd.	Name and address	Items ordered	Price
Ship'd.	Shipping station	Conditions	Date paid

Schedule No. X—Purchases

YEAR ENDING ________________ 19___.

PURCHASES

DATE	STOCK PURCHASED		EGGS PURCHASED	
	No. and variety	Cost	No. and variety	Cost
Totals:				

Schedule No. XI—Miscellaneous Income

YEAR ENDING ________________ 19___.

ITEMS	DATA	TOTAL
Insurance..		
Sale of equipment....................................		
Show premiums..		
Feathers..		
Fertilizer..		
Other receipts..		
Totals..		

Schedule No. XII—Miscellaneous Expenses

YEAR ENDING ________________ 19___.

ITEMS	DATA	TOTAL
Repairs..		
Insurance..		
Interest..		
Taxes..		
Show expenses..		
Other expenses..		
Totals..		

Schedule No. XIII—Depreciation

YEAR ENDING DEC. 31, 19___.

ITEMS	First Cost	Probable Life	Rate of Depreciation	Amount of Depreciation	Age	Value at End of Year
Laying and Breeding Houses.............						
Incubator House........						
Brooder House.........						
Colony House..........						
Etc..................						
Totals............						

Chapter XVI

Sundry Topics

IN our final chapter there are sundry topics that remain for consideration.

Duck Raising

The duck industry is one of growing proportions. On Long Island there are about 40 duck farms and the annual output in ducklings is more than 350,000. The Pekin duck has preference above all others for market purposes. Ducklings are sold at ten to twelve weeks of age, when their average weight is about six pounds.

The Pekin lays from 100 to 140 eggs in a season. Eggs for hatching are always in demand. Eggs can be hatched either with a hen or in an incubator. Successful incubation depends upon an even temperature with adequate moisture and ventilation.

Ducklings should not be fed for 36 hours after hatching. They can then be started on stale bread, hard boiled egg, milk and sharp sand, making a crumbly mash. After the fourth day from hatching, feed a mash as follows:

Bran	3 parts
Wheat Middlings	2 parts
Corn meal	3 parts
Meat scrap	5 parts
Sand or grit	2 parts
Green feed	5 parts

These ingredients are determined by measure and should be made into a moist mash and fed four times a day. Water must be supplied in abundance. During the winter season the breeding stock should be fed whole grains and should have access to grit and water.

Ducks very rarely become sick and for that reason are easily raised. Some of the more common diseases are cholera, coccidiosis, enteritis and worms.

Feathers are of considerable value. Unless the producer is skilled in plucking, it will be more profitable to ship alive than to dress for market.

Toulouse Geese

Raising Geese

Geese are profitable for flesh, eggs and feathers. A goose lays about 40 eggs in a season. When they are sold for hatching the price averages about 50 cents each. Incubation is usually accomplished by a chicken hen, which makes a good mother and brooder for the goslings until they are able to shift for themselves. For the first two weeks the goslings should be fed a mash of equal parts of corn meal, bran and wheat middlings, and about ten per cent of meat scrap with coarse sand for grit. They should have range for green feed from the beginning, and after they are old enough to forage they will need but little feeding, as grass and weeds make the bulk of their ration. An abundance of water is always necessary. Geese and chickens do not thrive well together on account of the unclean habits of the former and their pugnacious disposition. Many a choice cockerel has suffered a broken limb or ugly rents in the skin because of their vicious nature. If kept in separate runs or fed on separate grounds some of the objections might disappear.

It is very difficult, if not impossible, to detect the sex of goslings. When mature, however, it is found that the gander has a sharp, shrill voice, while the goose has a coarse, heavy voice. The male has a heavier, longer neck, and a larger head.

During the breeding season the gander can be detected by his actions. Another method is to use slight pressure upon the sphincter muscle in the region of the anus, when the sexual organs will be everted.

Geese very seldom succumb to disease, but occasionally are afflicted with cholera, coccidiosis, gout, rheumatism, or liver disease.

The feathers of geese are very valuable, commanding about 50 cents per pound. It is not a common practice, nor is it usually safe, to pluck geese alive. In dressing for market they should be plucked dry, and the fine feathers are always saved. There is little demand for the coarse feathers.

The leading varieties are the Toulouse and the Embden.

In fattening for market, ground corn, bran and 10 per cent of meat scrap make a good fattening ration. Geese should be mated in the fall. The loss of a mate is followed by a season of grief. A new mate is not always kindly received.

Turkey Culture

Turkey culture is both interesting and profitable. Eggs for hatching, feathers, and market birds are always in demand. Eggs for hatching sell for 30 cents to $1 each. Market birds for Thanksgiving and the holidays command from 30 to 50 cents per pound.

In mating turkeys it is customary to introduce new blood every other year. This is a good policy if care is exercised in selecting stock that is free from any taint of disease. Two toms should be provided, even for a small flock, one yearling tom and one young tom. The unexpected frequently happens, and the loss of a tom during the breeding season is not easily replaced. Yearling hens make the best breeders, but it is always advisable to keep as many well matured pullets as hens. In-breeding is considered harmful to vigor and vitality. By using old toms with pullets and young toms with hens I maintained a flock of turkeys for nearly ten years without the introduction of new blood and could not observe any decline in vigor or size.

The mating season in the northern states usually begins in March, and the hens begin laying early in April. One copulation fertilizes all the eggs of a clutch. A turkey hen lays about 40 eggs in a season, if not allowed to sit, in three clutches—18 eggs in the first clutch; 12 in the second, and 10 in the third. Eggs laid early in the spring should be gathered daily to prevent chilling. Suitable nests should be provided in barrels, boxes or brush piles, and if hens can be induced to begin laying at home there will be little danger of their roaming to find nests.

The turkey hen makes the best incubator and brooder. It is customary to hatch the first laid eggs under a chicken hen. In that event the poults should receive the same care and feed as chicks. As soon as a turkey hen hatches a brood, the hen-hatched poults can be transferred to the turkey mother. If

This will keep her from flying the fence

the weather is damp and cold at the time a brood of poults is taken off, it is best to confine the turkey mother in a coop for a few days. The coop should have an outside runway to give the poults opportunity for exercise and to obtain green food. Rolled oats, fed sparingly, three to five times a day, makes a good starter. This, with sweet milk and sharp sand for grit, will be all that is needed for the first week. After a week of confinement the hen with her brood can be given free range. They can be taught to come home for food and shelter each night, and as a rule this is advisable. Turkeys and chickens should not be fed on the same ground. In the first place, the turkeys are likely to infect the ground with gape worms; and, in the second place, they are in danger of becoming infected with certain diseases, such as coccidiosis, from the chickens.

In fattening turkeys for market it is customary to feed a mixture of grains, such as corn, wheat, and oats. If fed an excess of new corn, diarrhea is likely to occur, and this may lead to serious disease. If not shipped alive, turkeys should be dry-picked, plumped by plunging into cold water, and shipped undrawn.

The diseases which attack turkeys are cholera, blackhead, coccidiosis, limberneck, chicken pox, roup and intestinal worms. The treatment of these diseases is given in Chapter XIV.

Guinea Fowl

The number of guinea fowl on farms on Jan. 1, 1920, was 2,410,421, and the value of these was $1,582,313.

The system of feeding guineas resembles that of chickens, but they eat less food. Bread crumbs and hard-boiled egg make a good feed for the guinea chicks.

It is about as difficult to determine the sex of guineas as of geese. The male has a longer helmet and wattles and coarser features. The cry of the female resembles the word "buck-wheat," while that of the male is a one-syllable shriek.

Quoting from Farmers' Bulletin, No. 858:

"Guinea fowl are growing in favor as a substitute for game birds, with the result that guinea raising is becoming more profitable.

"They are raised usually in small flocks on general farms, and need a large range for best results.

"Domesticated guinea fowl are of three varieties: Pearl, White and Lavender, of which the Pearl is by far the most popular. Guinea fowl have a tendency to mate in pairs, but one male may be mated successfully with three or four females.

"Guinea hens usually begin to lay in April or May and will lay 20 to 30 eggs before becoming broody. If not allowed to sit they will continue to lay throughout the summer, laying from 40 to 60 or more eggs. Eggs may be removed from the nest when the guinea hen is not sitting, but two or more eggs must be left in the nest.

"Ordinary hens are used commonly to hatch and rear guinea chicks. The period of incubation is 28 days. Guineas are marketed late in the summer, when they weigh one to one and one-half pounds at about two and one-half months of age, and also throughout the fall when the demand is for heavier birds."

Judging

There are two methods of judging, by **score-card** and by **comparison.** The score-card is usually used in the smaller shows and is a source of information to the breeder. If his bird is disqualified or cut in any section he knows the fact and generally the reason why.

The comparison method teaches him nothing, except that he was a loser or that he won a ribbon, unless perchance he should meet the judge who may take the pains to point out the defects in his birds. There is a strong tendency to drift away from the score-card, even in the smaller shows, on account of the excessive amount of work that is entailed upon the judge and secretary of the show. As an educational program for the new beginner comparison judging is of little value. He neither learns why he lost or why his competitor won.

Exhibiting

Every farmer with a pure-bred flock will be benefited by entering a few birds at the county fair or in the local poultry show. A knowledge of the weak points is worth while, as it may incite to greater endeavor.

Conditioning for the show requires considerable attention.

The show specimens should be caught and cooped several days before entry. The legs and feet should be cleaned by washing with soap and water. Use a stiff brush. Rinse and dry and then with a toothpick or a sharpened stick remove all dirt from under the scales. Then apply an ointment of glycerine and alcohol, half and half. A numbered leg band should be attached to the right foot below the spike. Examine each specimen for disqualifications. White specimens should be washed. First wet the feathers to the skin with tepid water, then apply soap, working it into the feathers; then rinse and use soap again. The second rinsing should be carefully done so as to remove every trace of soap. Dry in a warm room and allow the specimen to cool gradually to prevent catching cold.

After the birds have been returned from the show they should be kept in quarantine for a week or ten days to detect

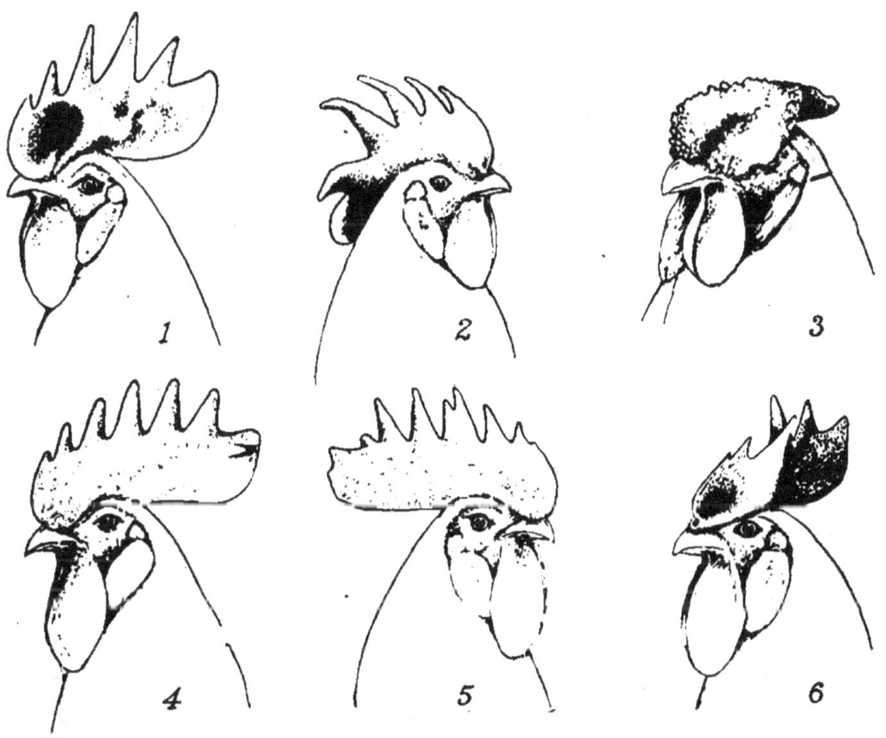

MALE HEADS SHOWING DEFECTIVE COMBS

1 THUMB MARK — 2 LOPPED (SINGLE) — 3 HOLLOW CENTER
4 SIDE SPRIG — 5 UNEVEN SERRATIONS — 6 TWISTED

the appearance of any diseases that may have been contracted in the show room.

Be a good loser and a grateful winner.

The Poultry Show. The poultry show has done more than any other institution, except the American Poultry Association, to perpetuate an interest in pure-bred poultry. It should be encouraged in every community. In selecting a place to exhibit the worth of his birds, the breeder should not neglect the local show. If victorious there, and he would have other fields to conquer, he should not forget the annual show of his State Poultry Association. This is his own organization and he should take advantage of the opportunities it proffers.

State Poultry Association. The State Association is affiliated with the American Poultry Association and it is also fostered by the State. Membership in this organization is always welcomed and should be counted a privilege and duty by every breeder.

Poultry Schools. There are several well organized and high-grade poultry schools in the United States. They are sponsored by capable and well-trained experts, and no mistake can be made in enrolling for a special course of study in one of these schools. It is knowledge that counts, and its application makes for success. Ignorance is the way of failure.

The Farm Bureau. This organization seeks to foster every farm enterprise. It is accomplishing much good along poultry lines. Get in touch with your farm adviser and he will gladly help solve some of your poultry problems.

State and University experiment stations are accomplishing a work for poultry culture equal to that of any other agency.

Egg Laying Contests. These are being conducted in some states under the direction of the State Poultry Association. Sometimes they are fostered by the state or by the state experiment station, again by private enterprise. The facts discovered and presented to the world by these contests have given a great impetus to the industry. These facts bear upon the problems of feeding, breeding, culling and production and enter into the practical problems that are met on the farm.

Poultry publications. The farm paper and poultry journals are the means of disseminating knowledge. They are working continually for better poultry and more of it. To remove these potent influences for good would be to set back the industry for many years.

The Poultry Calendar

January. This is the month of severe cold. Every precaution should be taken that the fowls receive proper protection. On severe days they should be confined to the poultry house.

Mate up the fowls that are intended for flock-breeding and give them special care. They should not have a forcing ration, should not be overcrowded, should not be kept under artificial lights, should be compelled to exercise for their food.

Clean up and disinfect the poultry house, replenish nesting material and litter in the scratch rooms.

Make an inventory of stock and equipment.

Begin a systematic method of poultry accounting. See Chapter XV.

Get the incubator ready for early hatching.

February. The egg yield is now increasing rapidly. A few of the breeding stock hens may become broody, but if early hatched pullets are desired, the incubator is the main reliance. Set the incubator early in February.

Clean up and disinfect the poultry house, supplying new litter and nest material.

Go over the breeding pens and breeding flock carefully to detect and remove any birds lacking in vigor.

Get the brooder house and brooder ready for the forthcoming hatch. If artificial lights are used and there is more than a 50 per cent yield, the lights should be reduced.

Gather eggs twice daily in severe weather.

Fatten and market all capons held over.

March. Clean up and disinfect the poultry building, supplying fresh litter.

Spray roosts and nests with some good lice killer. Dust the fresh nesting material with insect powder.

Give the laying flock and the breeding flock a thorough culling. Condition and ship to market all culls. This is the month of high prices for market stock.

Carefully examine the breeders and replace any male birds that are lacking in vigor or that fail to produce fertile eggs.

Prepare chick rations for the coming broods.

Bring the poultry account down to date and prepare the quarterly summary.

April. This is the busy month. Incubators, sitting hens, and tender chicks require constant attention.

Clean up and disinfect all coops and buildings.

Put on a vigorous campaign against lice and mites. Thorough work now will save future trouble.

Plant mangel-wurzels, beets, carrots, cabbage, turnips and other vegetables to supply the winter ration of green feed.

May. Clean and disinfect buildings and coops.

Clean, disinfect and put away the incubators, as the season for profitable incubation has passed.

All eggs set this month should be placed under hens. Hens should be protected from lice and mites and provided with new nesting material as required.

Provide shade for the chicks. Give the flock free range.

Give special attention to the chicks. Keep their quarters clean and free from drafts. Protect from enemies by closing them up at night and keeping the hen confined in the day time.

June. Clean and disinfect all buildings and coops.

Break up sitting hens by putting them in a strange lot or room without nests or in a breaking coop.

Produce infertile eggs for market by separating all males from the hens. The undesirable males should be shipped to market but those that are to be kept over may be kept in a separate lot.

Cull the three-months-old chicks. The surplus cockerels can be caponized or shipped as broilers.

Care should be used in culling the hens, as some of the best layers may be taking a rest or be broody. There are always some that cannot be profitably retained, and these should go to market.

Do not forget shade for the growing chicks.

Provide a mash for the summer layers. See Ration No. VII.

Bring the poultry account down to date and prepare the quarterly summary.

July. This is usually the hottest month and growing chicks and fowls must have protection from the intense heat. The orchard or corn field makes a fine shady run for the chicks. Artificial shade should be provided where necessary.

Remove all young stock to colony houses or to permanent roosting quarters.

Provide a dry mash and green feed for the young stock.

Clean and disinfect all buildings, also the brooders, storing them for future use.

Avoid overcrowding of young stock.

Liberal feeding is recommended for this month to promote growth and vigor and fortify against weather conditions.

August. Clean and disinfect all buildings. Supply roosts for young stock; caponize surplus cockerels; avoid overcrowding; protect against weather changes; store eggs for the time of scarcity.

September. Clean and disinfect all buildings and grounds as far as possible. Sprinkle slaked lime on ground, plow or spade, then supply another coat of lime. Later sprinkle on the ground a strong solution of copperas.

Cull the whole flock and, after conditioning, ship the culls to market. Fatten surplus young stock for roasters.

Make the quarterly inventory, bring the poultry account down to date, and prepare the quarterly summary. This is done at the end of month.

Provide a suitable mash for the molting hens, Ration IX.

October. Clean and disinfect all buildings, and make all necessary repairs for the winter season.

Watch out for colds. Keep the young stock from crowding and becoming overheated, afterwards to become chilled by the cold morning air.

Put all stock in winter quarters.

Watch closely for the occurrence of disease and take prompt measures to prevent the spread of any epidemic.

Self-feeder for dry mash

November. The best layers will molt in November and October. The molting ration should be continued until the molt is practically finished.

Make all changes in rations gradually.

Do not force for egg production pullets that are not matured.

Clean and disinfect all buildings and provide fresh nesting material.

Dispose of surplus breeders.

Matured pullets should be put on a laying ration.

December. Clean and disinfect all buildings.

Select breeding stock for special pens and for the breeding flock. Arrange to keep these in separate quarters so that they may have special care in feeding.

Bring the poultry account down to date and prepare the quarterly and annual summaries.

Cull the flock and ship surplus stock to market.

Postscript

Our task is finished. It is a big subject. At the best we have only skirmished along the outskirts of a few fields, as yet but partially explored. We have written little that is new, but have drawn from practical experience wherever possible. To the army of investigators who have blazed the way in many fields belongs the credit for the accumulated facts pertaining to the industry. We have gathered a few data with the avowed object of being helpful to the farmer and the fancier. That there are errors is freely admitted. Poultry writings have many contradictions. Some must be in error, yet each has given his contribution toward the attainment of an exact science.

A last word. What is the most important factor in the poultry problem? What word expresses the condition which, probably more than any other, conserves the health of the flock, determines its productiveness, and contributes to the profit of the industry? Is it not "cleanliness"? And next to cleanliness is "exercise".

Appendix

THE following pages are devoted to articles reprinted from PRAIRIE FARMER. The stories of farm success with poultry are particularly interesting as examples of what other farmers are actually doing to make the poultry flock add substantially to the farm income.

When to Market Poultry

THE chart on Page 218 portrays the usual seasonal changes in chicken prices. The lines represent five-year averages by months, using the prices to producers in the United States as estimated by the Department of Agriculture in one case and Chicago prices for spring chickens and hens for the other two lines.

Up to the end of April, the quotations on springers at Chicago are upon chickens hatched in the previous spring. Beginning with May such chickens are counted as hens and stags and the springers or broilers quoted are from the new hatch. Since few are hatched in winter and extremely early spring these spring chickens are very scarce in May, June and July so that they bring a big premium over the general run. The premium is gradually lost as such birds become more abundant and quotations upon them in late fall and winter are practically the same as upon hens. During March the springers have sold higher than hens, although the reason for this is not clear, especially as they seem to sell again in the same notch during April.

It is noticeable that the prices of both springers and hens decline on the average during the late summer and fall to the lowest point of the year in November when receipts are largest. Holiday demand which develops late in November and is prominent in December causes December prices to average materially higher than November. Hens reach their highest point as a rule in April when they are laying heavily and are kept back on farms. From this point prices decline as supplies become more abundant. June shows up as a month of low prices for hens partly because of the fact that they are in

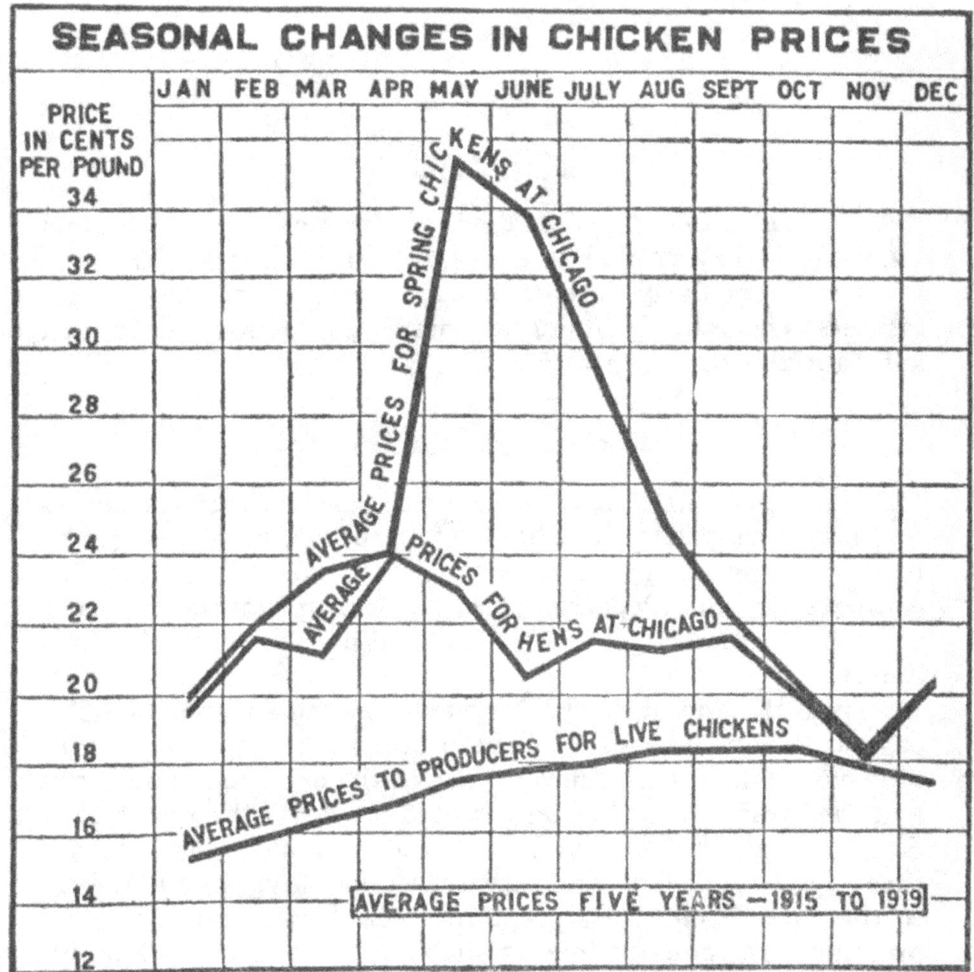

The Chicago prices are quotations at wholesale for live hens and spring chickens. Note that spring chickens are highest in May and decline rapidly during the summer to the low point in November, after which there is an advance during the winter months. Up to the end of April the spring chickens quoted are those hatched in the previous season. Hence the very sharp advance from April to May when the new hatch is quoted. Hens reach the highest point in April during the laying season. Thereafter they are marketed in larger numbers and June prices usually are low. Another decline takes place in the fall due partly to the abundance of spring chickens. Beginning in September and continuing through most of the winter hens and spring chickens fluctuate together.

Prices to producers are averages for the entire United States. They show less marked changes from month to month and are highest during the late summer and early fall months. They also seem to behave differently from Chicago prices.

poor condition when marketed immediately after the laying season.

The prices to producers shown are averages made up by the Department of Agriculture from the statements of a number of country buyers located throughout the United States

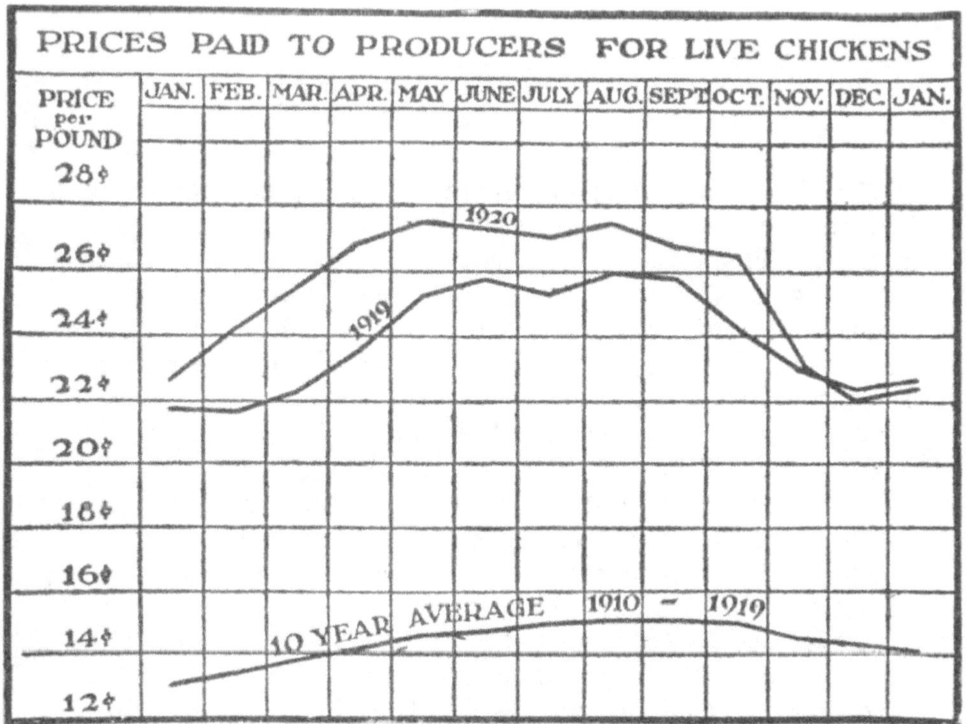

The prices to producers shown on the chart above are averages for the entire United States based on the reports for the first of the month. They are highest during the summer months. Note that the ten year average line is similar in its course to the five year average on the other chart. Prices in 1919 and in 1920 are similar to each other in their behavior and somewhat similar to the average for ten years except that they are on a much higher level.

who report the prices paid at country buying points. The curve for these prices is not consistent with the curve for Chicago prices for reasons not entirely apparent. Prices to producers have averaged highest during late summer and early fall and are lowest during the winter and early spring months.

To sell best on the open market poultry should arrive from Tuesday to Friday. Demand usually is light on Monday and on Saturday prices are often reduced so as to effect a clearance and avoid holding stock over till Monday

Poultry Marketing Experience

MANY Prairie Farmer readers have added materially to their poultry income by finding markets for their poultry and eggs which pay more than they can secure by selling to the village storekeeper or huckster. One Illinois farm woman, for example, netted 60 cents per dozen by shipping eggs to a New York commission house when eggs were bringing but 40 cents at home.

The increased returns on that transaction were 50 per cent, but the increased profit was much more. The cost of producing those eggs was 36 cents, according to figures kept by this woman. The profit was four cents per dozen when the eggs were sold locally, while the other method of marketing gave her a profit of 24 cents.

This case is not unusual. Rather, it is typical of what may be done by any farm poultry raiser in the Corn Belt. The margin is not always as much as in the case indicated, but on the other hand it is often much more. The margin secured from improved marketing may increase substantially the poultry profit, or it may mean the difference between a good profit and a discouraging loss.

In the December 3, 1921, issue of Prairie Farmer readers who have found methods of increasing their revenue from poultry by better methods of marketing were asked to write us their experiences. The letters received indicate that there are no less than 10 practical plans which may be followed by those who want better prices for their poultry products. They are:

1. Selling eggs to private customers.
2. Selling eggs to city markets.
3. Selling eggs for hatching.
4. Selling eggs to hatcheries.
5. Producing late fall and winter eggs.
6. Selling early fries.
7. Selling baby chicks.
8. Selling dressed poultry.
9. Shipping live poultry to city markets.
10. Selling cockerels and pullets for breeding purposes.

High prices for poultry and eggs mean that the quality must be first-class. The Chicago man who pays a ten-cent premium for fresh eggs from the country will refuse to pay that premium if the eggs are not strictly fresh, clean, uniform in size, and first-class in every way. Live poultry won't command the top prices if it isn't fat, and dressed poultry must be fat, clean and attractive in order to bring a high price. Quality goes hand in hand with better methods of marketing, but in no product is it more important than poultry and eggs.

Private Customers

Many farm women have secured better than market prices by selling eggs to private customers. In small towns the margin paid over store prices is often not more than two or three cents per dozen. The development of the parcel post has extended the private customer field to distant cities, and with the city trade has come a much better premium over the local egg markets.

Mrs. Chas. Plondke of Grant county, Wis., charges private customers five cents more per dozen than the local market, and the customer pays all expense of shipping the eggs and returning the empty cases.

Mrs. Clara Elfrink of Cook county, Ill., has 14 customers in Chicago who pay a premium of several cents per dozen for eggs. "We got our first customer nine years ago through the Chicago postoffice, and the others were friends of the first customers," she writes. "We ship in egg cartons, using the two, three, and four dozen sizes, according to the customer's needs. We ship mostly by parcel post, but occasionally send large shipments by express. We are only 30 miles from Chicago, and in good weather our customers often drive out for their eggs, dressed poultry, vegetables and fruit. When we ship, the customers pay the postage and insurance. We have had no trouble in regard to pay, as all have been prompt."

Mrs. Otto Schulz of McHenry county, Ill., ships eggs in 12-dozen crates to Chicago families, several families dividing a case when one can't use that many eggs. "I have followed this plan for three years and am well pleased," she writes. "I get a couple of cents above market price, and my customers pay the parcel post charges. I am also selling dressed geese, ducks, young roosters and feathers to the same customers, charging less than they can buy them for and at the same time giving me more than I could secure at the butcher's or the store. Chicago people are glad to pay well for fresh eggs and poultry."

Another plan is followed by Mrs. C. D. Finkbeiner of Lenawee county, Mich., who lives 50 miles or so from Detroit. Many automobile parties go past the farm, and a "Fresh eggs" sign stops enough automobiles to enable Mrs. Finkbeiner to sell all the surplus eggs except in the hatching season.

It is not difficult to make connections with city families who will pay a good premium for quality eggs shipped by parcel post. Most people have friends and relatives in cities, and that should enable them to get started shipping eggs. One satisfied customer brings another. In the larger cities the post-offices have lists of city people who want parcels post eggs. Cartons for shipping can be secured from supply houses, and may be secured at almost any store. Bills should be rendered the first of every month.

Shipping to Commission Houses

Shipping eggs to commission houses in the larger cities, as New York and Chicago, is a plan which nets a good premium over the local market. C. H. Poland of Montgomery county, Ill., in 1921, shipped all the surplus eggs from his 265 hens to a New York commission house and secured $185 more for them than his home market would pay.

A Missouri woman reports that her first shipment to New York netted 22 cents per dozen more than she could have secured at home, and that on the nine cases she shipped she netted $37.50 over local prices. These were all white eggs, which bring a good premium over other colors on the New York and Boston markets.

Mrs. Ed. Fulford of Jackson county, Ill., reports much higher prices secured in St. Louis than she could obtain at local stores. "I use a 12-dozen case and ship by insured parcel post," she says. "It costs 35 cents to send it when full and eight cents when empty. The eggs sell higher than at home and I can use the cash any place instead of taking what I can get, as I do at home. I ship the eggs one morning, get the check the next morning and the empty cases the following day."

Another who has found it profitable to ship to city markets is Merle Naugle of White county, Ind. "On November 22 I received $24.09 net for a 30-dozen case in New York, and the price here was $16.50," she says. "These were uniform white eggs which are higher than any others. In spring and summer it does not pay as well to ship east, so I sell to a man in Chicago who pays 14 cents over local prices."

The important things to remember in shipping to commission companies are to send uniform good-quality eggs, and to pick out a reliable commission man. There are some commission houses which are not trustworthy, but they are being rapidly weeded out by the state commission merchant laws. The premium over local prices is rather small from March to June, but the rest of the year it is considerable.

Eggs for Hatching

Those with well-selected purebred flocks (not necessarily fancy) find a profitable market as a rule in selling eggs for hatching purposes. This is true of all the common breeds like Leghorns, Wyandottes, Rhode Island Reds and Plymouth Rocks. Mrs. Lee E. McElroy of Shelby county, Ill., whose experience is typical of many, last year sold eggs for hatching at $1.50 per 15, selling all the extra eggs in that way from March till June. "The best answer to the question of better marketing of poultry is better poultry," she says. "No matter what market you sell to, purebreds return a good profit over scrubs."

Mrs. Frank Myers of Henderson county, Ill., began 10 years ago with a flock of Barred Rocks. "I culled my flock and bought the best stock I could afford," she writes. "By 1919 I thought my flock was good enough to advertise eggs for sale. Ads in Prairie Farmer and a local paper cost $10.09 in all. Baskets cost $2.70. Excelsior was free at the store, and flour sacks made the covers. I sold $92.50 worth of eggs. They were worth $48.30 at the store, advertising and other costs amounted to $13.40, so I had for my trouble $31.01. Since then I have advertised eggs, baby chicks and cockerels, Prairie Farmer and a local paper doing the work."

It is not difficult to secure $1 to $1.50 per sitting of eggs from well-selected hens, and eggs from flocks somewhat fancier than on most farms bring considerably higher prices. The expense is very small, and the returns usually are more than ample to pay for the time and trouble.

Eggs to Hatcheries

Since the rapid growth of the baby chick business, many farm poultry raisers with purebred flocks find that the baby chick hatcheries provide an excellent market for eggs. This plan eliminates many of the costs involved in selling hatching eggs, though it does not bring the returns which the other plan does. Mrs. Wm. A. Klein of Putnam county, Ill., has been selling Barred Rock eggs to a nearby hatchery, receiving several cents per dozen above market price. Another who has followed this practice is Mrs. Wm. M. Rhine of Montgomery county, Ill., who received five cents each for the eggs.

This plan offers considerable possibility to those who are near to hatcheries, and those whose flock is not large enough to justify advertising the eggs for sale. It is doubtful, however, if the hatchery trade is worth going after if one has a good flock for which a reputation is already established.

Producing Winter Eggs

Egg prices are always much higher from October until the last of January than at any other season. The farm poultry raiser who has learned the trick of making the hens lay in this "off" season gets ample reward in liberal egg checks. It is at this season, that city trade will pay the largest premium for fresh eggs, due to the difficulty of securing them and to the storage eggs that reach the market then.

The essential points in securing eggs in the months of high prices are proper feeding, proper housing and a productive flock. Complete discussion of this matter will be found elsewhere in this volume.

Selling Baby Chicks

The baby chick business is pretty strongly established, and offers a fine chance of financial reward to the person who has the right knack and patience. "I sold 900 Wyandotte baby chicks last spring at 15 cents apiece," says Mrs. W. H. Rigsby of Christian county, Ill. "They are easy to dispose of. Good, lively, healthy chicks are easily sold about as fast as they can be hatched."

Mrs. Arthur Heap of Kendall county, Ill., sold 843 baby chicks at 15 cents last spring, besides hatching 1,050 chicks for her own flock. "I sell them the day they are hatched, and there is very little risk," she writes. "I am satisfied that there is more money in the baby chick business than in any other branch of the poultry industry."

The baby chick business calls for a healthy purebred flock, and if managed on a modest scale does not call for any extra equipment. It is easily possible for a farm woman to sell 1,000 or more baby chicks in her immediate neighborhood without much expense. That sort of a business must be built up from a smaller beginning. The commercial baby chick business cannot be undertaken as a sideline to farming, but should be regarded as a highly specialized industry by itself.

Selling Early Fries

During the late spring and early summer months, young fries command high prices as a rule, especially if they are fat and plump. Mrs. Heap reports receiving 49 cents per pound for ten-week-old fries which averaged two pounds apiece. Others kept until they weighed five pounds apiece late in the fall were sold for 19 cents per pound or 95 cents each. The early chickens thus brought three cents apiece more and required less than half the feed and work.

"We sell most of our cockerels about July 1 as broilers," writes Mrs. Finkbeiner. "They then weigh two pounds each and command the highest prices of the year. To get them to that age we feed sour milk, cracked corn, oatmeal and bran."

Mrs. Wm. Klein of Putnam county, Ill., sold early fries the latter part of May that weighed two pounds and a quarter. Later when the price dropped below 35 cents, she quit selling, caponized the cockerels and sold the capons late in the fall for a fancy price. "There is practically no extra work in this plan," she writes, "and it is a very profitable marketing plan."

The early fry requires early hatching, and feeding for rapid gains, and if large numbers are produced a special market.

Selling Dressed Poultry

Some poultry raisers have gone a step further and eliminated the butcher by selling dressed poultry. "I believe that this is the best way to make extra money from poultry," says Mrs. B. C. Lawhead of Champaign county, Ill. "I make a profit of about 10 cents per pound on the average, for dressing the chickens. Before Thanksgiving, ducks were 22 cents alive and 30 cents dressed, geese 18 cents alive and 33 cents dressed. I cleared over $1 per goose and 35 cents per duck by dressing."

The experience of Mrs. Louis Bernhard of Effingham county, Ill., is similar. "Two weeks before Christmas we dressed 82 turkeys, averaging 15 pounds, for which we received 49 cents net after shipping to New York. We would have received 34 cents at our local stores. Thirty-five late turkeys were sold in February for 63 cents dressed (local price 34 cents). On these 117 turkeys we made $270 above local prices, after paying all expenses including shrinkage."

Mrs. Schulz, who has Chicago customers for eggs, finds that these same people like to get dressed chickens, ducks and geese for which they pay premiums over the market.

The dressed poultry trade is one which can be operated on as large or as small a scale as desired. The principal points to be remembered are to deliver the dressed poultry fresh and to use only fat, plump fowls.

Shipping Live Poultry

Shipping live poultry is largely a matter of finding a commission man who will handle the shipments properly and honestly—which is much less difficult than is generally supposed. In every large city there are many commission merchants who make a specialty of handling not only live poultry but also dressed poultry, veal, eggs, produce and a variety of other things. The net amount received after commission charges, express and other costs are paid is usually much higher than what could have been secured at local markets. PRAIRIE FARMER will help you find a reliable commission man, if you wish our assistance.

The Brown County (Ill.) Farm Bureau began in August 1921 to make cooperative shipments of poultry in connection with the livestock shipping association. The shipments were sent to city markets, and the proceeds, less than the transportation and commission, paid to the producers. Within a few months, 100,000 pounds had been handled, for which more than $6,000 above local prices were secured — an average of more than six cents per pound.

The Purebred Market

S. M. Phelps of Warren county, Ill., tells of a small poultry flock from which he made sales of cockerels and pullets of more than $100 per year for seven years—a record which is equalled by many other farm poultry raisers. The average return from this kind of marketing is from 50 to 100 per cent higher than selling on the market.

Anyone with a good flock can sell a number of cockerels in the home community without much effort, and as the flock becomes better established and more widely known the market for good cockerels is enlarged. There is not so much chance to sell pullets to the farm trade, as most people find it cheaper to buy eggs and hatch their own pullets.

These statements do not refer to the breeders with fancy strains. The business of producing the fancy strains of poultry is a specialized industry by itself and hardly comes under a discussion of marketing of farm poultry.

The methods described in this article are those found practical by farmers and farm women who run their poultry business purely as a sideline, though any sideline to be profitable must be efficiently handled. The improved methods of marketing poultry and eggs which are described here help promote that efficiency and make possible profits where losses existed before, and make greater profits where there were small profits.

The man who wins in any branch of farming during the years following the war will be the one who farms most efficiently. And one of the chief factors in efficient farming is marketing products to the best advantage.

No Eight-Hour Day for Esgar's Hens

WHEN W. J. Esgar of Grundy county, Ill., gets the fire built about half past five on a cold winter morning he hurries down to the chicken house to call the hens. The way he calls them is to turn on the electric lights. Unlike the hired man, the hens never turn over in bed for another wink. They get off the roosts at once and go after their breakfast of sprouted oats and wheat. They get in a good two hours' work before daylight, and by the time the late winter dawn is waking up chickens who go by sun time, Esgar's hens have their crops full and are ready to go to laying.

"The value of this extra two hours' work by electric light is that the hens have just that much more time to eat, and they lay in proportion to the amount of feed they consume," says Esgar. "When a hen is eating she is making you money. When she is sleeping, especially in the morning with an empty crop, she isn't."

Esgar is a firm believer in poultry as a profitable branch of the farm business. He doesn't believe in leaving all the work to mother, either. He and the five boys are as much interested in the poultry flock as mother is, and are always ready to lend a hand when there is work to be done around the poultry yards.

As for the actual figures—well, there was a check for $145.25 for 130 broilers (10 weeks old) sold last June (1921). Egg receipts for 1920 were $216.50, and about as many eggs were used at home as were sold.

Keeping Egg Records

Esgar has been using trap nests for 12 years. The trap nest is a nest with a little trap door in front that closes when the hen goes in to lay. She has to stay there until she is let out, when she is credited with her egg.

"No, it isn't so very much trouble," says Esgar. "We put numbered leg bands on the pullets, and after that it isn't much work to open the nests and put down the records. It is the only sure way to build up a heavy-laying strain."

He showed me some of the records and we added up a few of them. One pullet laid 92 eggs from Jan. 19 to May 23, 1921—124 days. At that rate her year's record will be well over 200 eggs. Hens like this are mated with roosters from high-producing strains. Esgar paid $15 last spring for a rooster whose mother had a record of 300 eggs a year.

The results of this work are rapid improvement of the laying ability of the flock. In June, 1921, the egg production was 90 to 100 eggs a day from 125 hens. Esgar has so much confidence in the laying ability of his birds that he recently sent a pen to Murphysboro to compete in the state egg-laying contest.

Hens Lay When Eggs are High

"Trap nesting adds greatly to the interest in poultry raising," says Mrs. Esgar. "We have all taken much more interest in the flock since we began keeping laying records."

Esgar's winter laying flock consists of about 150 hens and pullets—mostly pullets. They are usually getting well under way by the first of November, when eggs are getting up toward the high point of the year. Most of his flock are Rhode Island Reds, though he is experimenting with Rhode Island Whites.

"We start the incubators in February so as to sell the cockerels for early broilers," says Esgar. "This spring we sold them at 40 to 55 cents a pound when they were 10 weeks

This is Billy Esgar and one of his roosters

old. They averaged over two pounds and brought a little over a dollar apiece. The chief difficulty with early hatching is that the pullets mature so early that they are likely to molt the first year, which spoils the November laying. We can usually prevent that, however, by shutting off the feed after the cockerels are marketed, letting the pullets range for a living. That checks their development and in most cases prevents molting the first year."

Esgar has a large winter laying house with an open front. It did not cost much, but it is as warm and comfortable as the most expensive house built. The only openings are in front. The open front gives plenty of ventilation without drafts.

The winter bill of fare is as follows:

Morning—Whole oats scattered in the litter.
Noon—Sprouted oats and whole wheat.
Night—Cracked corn scattered in the litter.

As additional green feed the hens get all the mangels they want. Self-feeders are kept full of a dry mash of equal parts of bran, ground oats and ground corn, with 10 per cent by weight of tankage. The hens also get all the skim milk and water they want to drink, and oyster shell and charcoal are kept before them all the time.

The laying hens are kept in the house all winter.

"They won't lay if they have wet or cold feet," says Esgar. "A few hours walking around in the snow, or even on frosty ground, will cut the egg production severely."

The floor of the house is covered with dry straw, changed once a week or so. The dropping boards under the roosts are cleaned frequently, and roosts, nests and the house itself are disinfected often enough to keep them free from vermin.

"I don't know of any branch of farming that beats the poultry business, good years and bad years alike," Esgar says.

Chester Married the Right Girl

WHEN Chester Winsor came back from a year's heavy action in France he found that he had a still harder fight before him. Like many other farmer boys who gave their service to the cause of liberty, he started farming when everything was high, and was pretty thoroughly deflated by the time the bills came due.

But Chester married a girl who believes that a farmer's wife should be his partner in every sense of the word. While Chester is plowing corn and making hay on his 80-acre rented farm in Grundy county, Ill., she is raising chickens and hunting eggs.

What Do You Think of This?

While I was eating some of her fried chicken one day not long ago she got out her account book and figured up her poultry income. During the 10 months from Dec. 1, 1920, to Oct. 1, 1921, her cash receipts from poultry and eggs were $801.47. The actual cash expense was $125. This poultry income, which has been a mighty big help during the deflation year, was all from market stuff. Although Mrs. Winsor has a flock of purebred Rhode Island Reds, the poultry and eggs are sold on the market and not to the purebred trade.

The first essential to success with poultry, particularly in winter egg production, is a good poultry house. This is one of the obstacles in the way of a tenant farmer who starts out to increase his income by the poultry route. Chester Winsor solves this problem in a way that is open to almost any tenant farmer. He started with an old shed that was of little value. With a few dollars' worth of second-hand lumber he made it into a very good substitute for a high-priced poultry house. The completed building is about 20 feet deep, with an open front scratching shed to the south, and windows above where the scratching shed joins the main building to let sunlight on to the roosts. The outside of the building was covered with brown building paper and painted with tar paint. The completed house is good for several years, and is as warm and comfortable as any laying hen could ask for. A month or two of winter eggs will pay the entire cost.

Dirt Floor O. K.

The house is large enough for 100 hens, which is about the size of the winter laying flock. It has a dirt floor, which is as good as any, Chester says, if it is kept dry. A trench around the outside of the house carries away the surface water. The floor is kept well bedded with dry straw. The front of the house is open, covered only with wire netting. "It might be an advantage to have curtains to let down in zero weather," Chester says. "I haven't felt the need of them yet, however. There are no side or rear openings, so there is never any draft, and in the rear, where the fowls roost, it is always warm."

The Bill of Fare

Mrs. Winsor uses a prepared dry mash for her laying hens, and a prepared chick feed for the small chickens. The rest of the ration is home-grown feed—skim milk, corn, wheat, with sprouted oats and mangels for green feed.

"One of the biggest elements of success in the poultry business is a healthy flock." says Mrs. Winsor. "Proper care and feeding, and a poultry house that is free from drafts and vermin, are necessary to good health. I have a neighbor whose poultry crop is almost a failure this year in spite of good care. The trouble is low vitality caused by a siege of roup in the breeding flock last winter. She would have been money ahead if she had sold the whole flock last spring and bought baby chicks or hatching eggs."

Mrs. Winsor says there is no reason why any other tenant farmer's wife cannot do as well with poultry as she has done. All it takes is the knowledge of a few fundamental facts about care and feeding, plus a willingness to put that knowledge into practice, plus a hubby willing to spend a little spare time fixing up a poultry house. And if the farm woman can show a cash income from her flock like that secured by Mrs. Winsor, what husband wouldn't be willing to help her now and then?

Hens Helping to Pay for Farm

PAYING for a 240-acre Illinois farm is something of an undertaking. Ray Coop of Kendall county, Ill., realized that when he moved to his new farm the first of March, 1921. But he is cheered by the thought that he won't have to do the job alone. Mrs. Coop is one of the best poultry raisers in all Northern Illinois, and she and her chickens are doing effective work in helping pay for the farm.

The hens seem to realize their responsibility, too, for they celebrated their first day on the new farm last spring by laying 56 eggs. When you stop to consider that there were only 99 of them, and that they had been moved several miles over rough roads, you will have to admit that they started out the new season with a pretty fair day's work.

Celebrating New Year's

The year's work really started on New Year's day, however, when Mrs. Coop put into effect her resolution to make 1921 a record-breaking poultry year by setting her incubator. The cockerels from that hatch sold as broilers in April at 72 cents a pound. They paid over $2 a bushel for the corn they ate, according to Mrs. Coop.

When I visited Mrs. Coop Oct. 26, 1921, she had already sold over $600 worth of poultry that year and had 50 more cockerels shut up ready for market. And that is saying nothing about the eggs. The egg checks for January and February alone came to, $150. So it looks as if Mrs. Coop would make good her resolution and exceed her 1920 record of $1,000 from her poultry flock.

A Good Business Woman

A good deal of Mrs. Coop's success is due to the fact that she is a good business woman as well as a good poultry raiser. She believes in having something to sell when prices are good. That is why she raises so many early broilers. When eggs went down to 18 cents last spring she stopped selling and set her incubator, selling the eggs as baby chicks. A little later she went to Joliet and made a contract with a large restaurant, which netted her a price at times as high as 15 cents above the market.

Her main reliance, however, aside from the early broilers, is winter eggs. She has 150 White Wyandotte pullets about ready to get under way for a heavy winter's work.

They have a deep poultry house that gives them plenty of room for exercise. The front part is open to let in the fresh air. A small yard will be built on the south side so that they can get outside when the weather is good.

The dry mash, which is fed in self-feeders, is bought ready made. Everything else is raised on the farm. The hens have plenty of skim milk, and this, by the way, is a standard feed for most of the successful poultry flocks in Grundy county. Instead of giving her hens milk to drink, however, Mrs. Coop waits until it is sour and thick before feeding it.

The scratch feed consists of wheat and cracked corn. Sprouted oats furnish the green feed that the hens need to keep them in good laying condition.

Mrs. Coop is an expert poultry culler, and she keeps the loafer hens sorted out of the laying flock.

"I love my hens and enjoy nothing so much as taking care of them," says Mrs. Coop enthusiastically, and that enthusiasm is one of the secrets of her success.

But then, who wouldn't be enthusiastic over a flock of hens that is doing so much toward paying for the farm?

Hens Help Pay the Bills

VERNE ANDERSON of Grundy county, Ill., has eaten three square meals a day all summer. He hasn't had to go in debt, either, thanks to his flock of hens and his two Hampshire sows and his four cows.

"I didn't know how I was going to make it in October, with the hens molting and the cows dry," he told me one day the latter part of that month. "But the pullets are beginning to lay now and the cows are coming in, so it looks like a pretty good winter."

Verne is farming 80 acres, which doesn't raise enough corn and oats to make much of an income these days. But when you feed that corn and oats to chickens and pigs you have another story.

"Chickens and eggs and butter money came in steadily all summer," Verne says, "and it was surely a big help. A few hogs now and then helped out, too. I didn't have enough of them, but I'm going to have more next year."

Anderson sold 226 dozen eggs from his flock of 110 white Wyandotte hens from Jan. 1 to Apr. 9. During that period he used 18 dozen in the incubator, besides eating unrecorded dozens. The price of these eggs ranged from 65 cents down to 18, most of them bringing 40 to 45 cents.

The Hen's Menu

The following is the daily menu of Anderson's flock of 100 or so laying hens: Eight pounds corn, five pounds oats, three pounds middlings, three pounds bran, and 1½ pounds tankage. The middlings, bran and tankage are mixed together to form a dry mash. The oats are fed whole and the corn cracked; both are fed in the straw so the hens have to work to get them.

"I like tankage for laying hens," Anderson says. "It is just as good as the more expensive beef scrap, and supplies the protein and animal matter that the hens need. Some farmers rely on skim milk instead of tankage, but my experience is that milk will not take the place of tankage. I feed some skim milk—from 3 to 3½ gallons a day—but even then it pays to feed tankage too.

"You will notice that my ration is made up of cheap feeds, mostly produced at home, so my feed bill is small. I don't even buy my middlings and bran, but get it from the mill when I get my winter's flour ground. I mix my mash feed very carefully by weight. I used to do it by guess and found it didn't pay.

Anderson has a modern poultry house 20 by 20 feet, built according to plans furnished by the farm bureau. "This poultry house alone has much more than paid my farm bureau dues," he says.

Verne Anderson's poultry house

Comfortable Winter Quarters

The house is the half monitor type. The scratching shed on the south has an open front. Anderson has curtains to cover this opening but never uses them. "You can't have too much fresh air if you get it without a draft," he says.

Above the scratching shed, on the south side of the main part of the building, are windows which let sunlight in on the roosts, and which can be opened for additional ventilation. The roosts extend across the back part of the house where there is never any draft. Roosts and nests are movable and can be taken out for cleaning and disinfection. Just under the roosts are dropping boards. The droppings can be easily scraped off into a wheelbarrow with a hoe, and the space underneath gives that much more room for the fowls to move around when they are shut in in bad weather.

The floor of the house is made of hollow building tile—seconds — laid flat and painted on top with a thin coat of cement. This makes a relatively cheap floor, rat proof, dry and easy to clean.

The materials for the house cost about $200. Anderson did the work of putting it up himself.

Keep the Hens Working

"No one can produce winter eggs without the right kind of a poultry house," says Anderson. "A house like mine is just right for 100 laying hens. They are comfortable all winter, and it is little work to take care of them. Such a house will soon pay for itself. I never could see the idea of keeping a flock of hens doing nothing all winter when eggs are at the highest price of the year. Proper shelter, a little care, and the right kind of feed, will keep them working at a profit all winter."

One big advantage of poultry on the farm, according to Anderson, is that it provides a steady source of income. With poultry and butter money coming in every month it isn't necessary to go in debt for running expenses.

"Poultry properly taken care of is much more than a side issue," he says. "It ought to be much more widely appreciated as a substantial source of farm income. If it were, there wouldn't be so many farmers complaining about hard times. A flock of good hens is the best hard times friend a man ever had.

Poultry Pays the Rent

IF I had a fine poultry house and modern equipment I could make money out of poultry, too.

How many times have you heard that remark? But wishing will never make the farm flock pay. If you live on a rented farm, that is all the more reason why you should make poultry add substantially to the farm income, for you need the money even more than the man who owns his farm and doesn't have to pay rent.

Let's Go Anyway

No equipment?

Well, you remember the story of the railway superintendent who was on the way to an important meeting when he was stopped by a washout. He wired the president: "Washout on line. Can't come."

And the president wired back: "Buy a new shirt and come anyway."

So let's go anyway, whether we have everything just as we would like or not.

That's what Mr. and Mrs. G. E. Holoch of Ford county, Ill., did last year, and their cash income from poultry was $802.84. They live on a rented farm, too.

"We do not have anything very up-to-date, but we make the best of what we have," says Holoch. "We felt that we couldn't afford to sit around and wait until we have everything just the way we want it before trying to make money from our poultry."

So they went at it, and made enough from the poultry to pay the rent last year.

How the Hens Are Fed

The Holochs raise a laying strain of White Wyandottes. For breakfast the hens get a hot mash composed of skim-milk and water, half and half, thickened with bran and alfalfa meal, with a pint of tankage added for each 100 hens. Dry oats are fed in the litter for a morning scratch feed, and ear corn is fed at night. Oyster shell and bran are kept in self-feeders all the time, and the water fountains are supplied with warm water.

They begin to set the incubator and hens the first of March. Last spring the egg market was not very satisfactory so after they had enough chicks for their own use, they hatched 1,200 for the neighbors.

The baby chicks brought $180, other poultry $364.88, and eggs $257.96, making a total of $802.84. The cash outlay for the year was not over $50. The feed, of course, was raised on the farm. It is certain, says Mrs. Holoch, that the grain fed to the poultry brought much greater returns than that marketed in any other way.

She says, too, that they are going to raise more poultry than ever this year, as they are convinced that there is nothing on the farm that pays better.

The poultry house is an old building, 12x30. In order to provide more light to supplement the four small windows they cut an opening three by six feet in the south side and covered it with a muslin curtain. This curtain provides ventilation, and when removed on sunshiny days, lets in lots of sunlight.

Feeding the Chicks

Mrs. Holoch's baby chick feed is composed of bread crumbs, ground corn and bran, mixed with sour milk and a little soda and salt and baked. They also get grit, and sour milk and fresh water to drink. They are fed nothing until they are 48 hours old.

Nothing so very difficult about that, is there? Then why not try it on your farm, and give the hens a chance to pay your rent for you?

How One Farm Woman Markets Poultry

IT IS necessary to use all the intelligence we have to market what we have to sell to the best advantage, says Mrs. W. A. McKeever, Ford county, Illinois. This applies especially to sidelines such as extra produce, butter, eggs, and poultry. A dollar or two here or there means more than it did when prices were more nearly fair to the farmer. The easiest way to sell most things is to exchange them for groceries or sell to local buyers, but it does not bring in the most money.

The poultry buyer is not just joy riding in his old Ford, nor is he gathering up chickens simply to accommodate his customers. He makes his living off the the profit he makes buying and selling. He seldom pays half the Chicago price for old cocks, nor within from five to 10 cents of the price on other stock. Why not ship direct to a reliable firm and make from two to five dollars more on a coop? If you do not know any reliable firms, ask your farm bureau to recommend one to you.

If you raise purebred stock, there is little need to sell anything except culls on the market. There is a good steady demand for good quality breeding stock at a nice margin of profit above market price. The market is always close at hand through advertising. Some think it costs too much to advertise. If it did not pay why would so many millions be spent on it every year? It always pays if you know how to fit your ad to your goods. The local paper is an inexpensive medium but it is only suitable for handling a small number because of its limited circulation. For a considerable number, 20 or more, it's hard to beat a good general farm paper with a strong poultry department.

I usually write out all that seems necessary for describing my stuff and then condense it carefully. Some start out "For

Sale." I think that is unnecessary. No one would think you were telling the world what you had, just to pass out information. Others use the terms "Full Blood," or worse yet "Thoroughbred,'" meaning "Purebred." It is not even necessary to pay for saying "Purebred." No one advertises scrub or grade poultry.

"Satisfaction Guaranteed" is another catch word that means little. If a customer is really dissatisfied, it is best to refund purchase price without argument. That's the way big mail order houses do, and it is their best advertising point. One disgruntled woman can do a lot of talking. I know for I have done it myself. When a certain commission firm gave me 25 pounds shrink on one coop of chickens, I certainly did not keep the news to myself, and I always think that my customers may be able to talk as much and as fast as I can. Of course no one ought to expect a bird that will win sweepstakes at the National Poultry Show for $5 but it is just as well not to promise perfection. I always prefer to send something just a little better than what is expected, if I can.

Going back to writing the advertisement it seems better to put a price in the first ad. I have tried both ways, and I know that an ad with a price in it gets more business. It scares off those who would not pay your price anyway, and saves the trouble and expense of answering their letters. Finally, a good clear letter often helps to finish what the advertising starts. A typewritten letter is preferable, and a letterhead also adds to the appearance of a letter. Most of your customers have only your letter to judge you and your stuff by, and the more businesslike it appears, the better impression it will make.

Timeliness is important in advertising. Most people are not ready to buy breeding stock, except hens or pullets, before Christmas, though there are always some forehanded enough to buy early in the winter so as to get first choice. The sharpest demand for cockerels and turkey toms comes in February, but I usually start to advertise after Christmas if I have many to sell. The heaviest demand for hatching eggs is in April, though some orders will come in March and some as late as May. Turkey eggs are in demand in May and early June.

Intelligent marketing through careful, consistent advertising undoubtedly pays even when poultry is only kept as a sideline, as it is on most Corn Belt farms. While the farmer's wife with her small flock of farm-bred, range-grown, chickens cannot compete with the professional fancier, she can have the satisfaction that comes through owning a nice flock, and have a few more dollars for Christmas, if she is willing to go to the extra trouble of selling her surplus good stock for breeding stock, instead of handing it over to the itinerant buyer. She cannot do it without some kind of advertising, and the right kind is by far the cheapest. Hiding your light under a bushel does not pay in these strenuous days.

Time Switch for Henhouse

HERE is a simple device for turning on electric lights in the poultry house in the morning, so as to get the hens to work early. Whenever the alarm goes off, the winding stem or key for the alarm begins to revolve and a string wrapped around this key or around a spool fastened to it will exert quite a strong pull, quite enough to operate the trigger which allows the knife switch to be pulled into place. In fact, the alarm should be wound only a turn or two so that too strong a pull

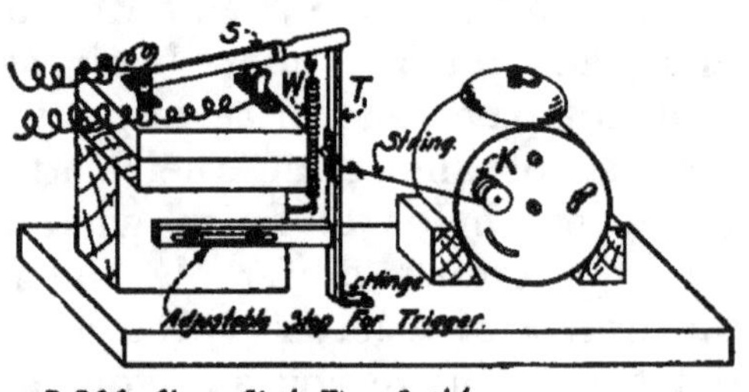

D-244. Alarm Clock Time Switch.

will not be developed and alarm run too long. The arrangement is shown in the diagram (D-244), S being the single-blade knife switch which turns on the feeding pen lights, T is the toggle trigger which holds the switch open, W the weight or spring which closes the switch, and K the alarm winding key with a string which pulls on the trigger when the alarm goes off. A trigger and weight are necessary, as the switch contacts will be burned unless the switch is closed quickly. The trigger should be of the toggle type; that will support the weight easily as long as it is straight, but which will double down whenever a small side pull is applied near the middle. Such a toggle could be made out of an ordinary folding foot rule.

Index

www.ingramcontent.com/pod-product-compliance
Lightning Source LLC
Chambersburg PA
CBHW081717220526
45468CB00008B/1884

* 9 7 8 1 7 2 1 8 4 1 0 8 0 *